Environment and Philosophy

Concern about the environment and what we are doing to it has put important questions on both moral and political agendas. One such question that is often asked in the West is whether or not we face a terminal environmental catastrophe in the foreseeable future? Less dramatic but equally serious threats to the sustaining powers and attractiveness of the world in which we live are also important considerations.

Environment and Philosophy provides an accessible introduction to the radical challenges that environmentalism poses to concepts which have become almost second nature in the modern world, including:

* the ideas of science and objectivity
* the conventional placement of the human being 'within' the environment
* the individualism of conventional 'Modern' thought

Written in an accessible style for those without a background in philosophy, this text examines ways of thinking about ourselves, nature, and our relationship with nature. It offers an introduction to the phenomenological perspective on environmental issues, and considers what natural beauty is and why threats to it are central to environmental policy decisions.

Vernon Pratt is a Senior Lecturer in the Department of Philosophy, University of Lancaster. **Jane Howarth** is a Senior Lecturer and **Emily Brady** is a Lecturer in the Department of Philosophy, University of Lancaster.

Routledge Introductions to Environment Series

Published and Forthcoming Titles

Titles under Series Editors:
Rita Gardner and Antoinette Mannion

Titles under Series Editor:
David Pepper

Environmental Science texts

Hydrological Systems
Oceanic Systems
Coastal Systems
Fluvial Systems
Soil Systems
Glacial Systems
Ecosystems
Landscape Systems

Environment and Society texts

Environment and Economics
Environment and Politics
Environment and Law
Environment and Philosophy
Environment and Planning
Environment and Social Theory
Environment and Political Theory
Business and Environment

Key Environmental Topics texts

Biodiversity and Conservation
Environmental Hazards
Natural Environmental Change
Environmental Monitoring
Climatic Change
Land Use and Abuse
Water Resources
Pollution
Waste and the Environment
Energy Resources
Agriculture
Wetland Environments
Energy, Society and Environment
Environmental Sustainability

Gender and Environment
Environment and Society
Tourism and Environment
Environmental Management
Environmental Values
Representations of the Environment
Environment and Health
Environmental Movements
History of Environmental Ideas
Environment and Technology
Environment and the City
Case Studies for Environmental
    Studies

Routledge Introductions to Environment Series

Environment and Philosophy

Vernon Pratt
with Jane Howarth
and Emily Brady

London and New York

First published 2000
by Routledge
11 New Fetter Lane, London EC4P 4EE

Simultaneously published in the USA and Canada
by Routledge
29 West 35th Street, New York, NY 10001

Routledge is an imprint of the Taylor & Francis Group

Typeset in Times by Taylor & Francis Books Ltd
Printed and bound in Great Britain by St Edmundsbury Press,
Bury St Edmunds, Suffolk

British Library Cataloguing in Publication Data
A catalogue record for this book is available from the British Lib

Library of Congress Cataloging in Publication Data
Pratt, Vernon, 1943–
Environment and Philosophy / Vernon Pratt with Jane Howarth a
Emily Brady
(Routledge Introductions to Environment Series)
1. Environmental sciences – philosophy. I. Howarth, Jane.
II. Brady, Emily. III. Title. IV. Series.
GE40.P73   1999   99-27647
304.2'01–dc21   CIP

ISBN 0–415–14510–4 (hbk)
ISBN 0–415–14511–2 (pbk)

For Ritsuko, Kencõ and Mari

Contents

Plates

Boxes

Series editor's preface
Environment and Society titles

The 1970s and early 1980s constituted a period of intense academic and popular interest in processes of environmental degradation: global, regional and local. However, it soon became increasingly clear that reversing such degradation would not be a purely technical and managerial matter. All the technical knowledge in the world does not necessarily lead societies to change environmentally damaging behaviour. Hence a critical understanding of socio-economic, political and cultural processes and structures has become, it is acknowledged, of central importance in approaching environmental problems. Over the past two decades in particular there has been a mushrooming of research and scholarship on the relationships between social sciences and humanities on the one hand and processes of environmental change on the other. This has lately been reflected in a proliferation of associated courses at undergraduate level.

At the same time, changes in higher education in Europe, which match earlier changes in America, Australasia and elsewhere, mean that an increasing number of such courses are being taught and studied within a framework offering maximum flexibility in the typical undergraduate programme: 'modular' courses or their equivalent.

The volumes in this series will mirror these changes. They will provide short, topic-centred texts on environmentally relevant areas, mainly within social sciences and humanities. They will reflect the fact that students will approach their subject matter from a great variety of different disciplinary backgrounds; not just within social sciences and humanities, but from physical and natural sciences too.

And those students may not be familiar with the background to the topic, they may or may not be going on to develop their interest in it, and they cannot automatically be thought of as being 'first-year level', or second- or third-year: they might not need to study the topic in any year of their course.

The authors and editors of this series are mainly established teachers in higher education. Finding that more traditional integrated environmental studies or specialised academic texts do not meet their requirements, they have increasingly met the new challenges caused by structural changes in education by writing their own course materials for their own students. These volumes represent, in modified form which all students can now share, the fruits of their labours.

To achieve the right mix of flexibility, depth and breadth, the volumes, like most modular courses themselves, are designed carefully to create maximum accessibility to readers from a variety of backgrounds. Each leads into its topic by giving adequate introduction, and each 'leads out' by pointing towards complexities and areas for further development and study. Indeed, much of the integrity and distinctiveness of the Environment and Society titles in the series will come through adopting a characteristic, though not inflexible, structure to the volumes. Each introduces the student to the real-world context of the topic, and the basic concepts and controversies in social science/humanities which are most relevant. The core of each volume explores the main issues. Data, case studies, overview diagrams, summary charts and self-check questions and exercises are some of the pedagogic devices that will be found. The last part of each volume will normally show how the themes and issues presented may become more complicated, presenting cognate issues and concepts needing to be explored to gain deeper understanding. Annotated reading lists are important here.

We hope that these concise volumes will provide sufficient depth to maintain the interest of students with relevant backgrounds, and also sketch basic concepts and map out the ground in a stimulating way for students to whom the whole area is new.

The Environment and Society titles in the series complement the Environmental Science titles which deal with natural science-based topics. Together this comprehensive range of volumes which make up the Routledge Introductions to Environment Series will provide modular and other students with an unparalleled range of perspectives on environmental issues, cross referencing where appropriate.

The main target readership is introductory level undergraduate students predominantly taking programmes of environmental modules. But we hope that the whole audience will be much wider, perhaps including second- and third-year undergraduates from many disciplines within the social sciences, science/technology and humanities, who might be taking occasional environmental courses. We also hope that sixth-form teachers and the wider public will use these volumes when they feel the need to obtain quick introductory coverage of the subject we present.

David Pepper
1997

Series International Advisory Board

Australasia: Dr P. Curson and Dr P. Mitchell, Macquarie University

North America: Professor L. Lewis, Clark University; Professor L. Rubinoff, Trent University

Europe: Professor P. Glasbergen, University of Utrecht; Professor van Dam-Mieras, Open University, The Netherlands

➊ Introduction

Awakening of concern over the environment and what we are doing to it has put large questions on the agenda. One that is often framed in the West is whether 'we' do indeed face a kind of *terminal* environmental catastrophe in the foreseeable future; though, of course, for large numbers of people the collapse of the systems they rely on is a present fact and not a future contingency. Less dramatic but still serious threats to the attractiveness and sustaining powers of the world we live in are noticed on all sides.

In so far as this state of affairs is a result of human decision-making, part of a thoughtful response is to try and develop that process so that it takes due account of likely environmental impacts. A field of discussion that has formed in recent years to contribute in this way is 'Environmental Ethics': it tries to work out what is properly to be understood as 'due' here. What 'environmental impacts' *should* be taken into account when we are deciding what to do? Why? How should different environmental considerations be weighted – *vis-à-vis* each other, and *vis-à-vis* considerations of other kinds? In addressing these questions Environmental Ethics has drawn on traditional discussions of ethics and the basis of ethics that have been developed within academic philosophy.

As those discussions are pursued, however, the focus of thought can shift. We may start by asking if we should allow ourselves to modify the genotype of a crop plant to make it invulnerable to disease, but such a question can easily lead us into questioning our presuppositions about what a living plant is, what a human being is, and how we conceive of the world we live in, and the little bits of that world which constitute our immediate surroundings. Environmental ethics leads to these questions and indeed addresses them, but they tend not to be its focus.

This book tries to concentrate on the presuppositions, which have been highlighted mostly of course by those who wish to subject them to challenge. Its theme is the adequacy of two great structures of Modern Western civilisation, individualism and science. These are the pillars that sustain the airy spaces of 'modern' life, it is often claimed – the life that pretty well has engulfed all others – and they are flawed, and failing.

First, conventional thought is said to be culpably 'individualistic' – that it thinks of human beings as rather like the balls on a billiard table, each sharply separated off from the others, bumping into each other occasionally, but not relating in any more complicated ways. There are two ways in which such a picture is inadequate, it is claimed. It gets wrong the relationship between one human being and another, and also the relationship between human beings and their environment.

Individualism (I will have to define it in more detail later) is usually thought to be characteristic in the West of the era ushered in by the rise of modern science. Often called the 'Modern' period, or 'Modernity', it is regarded as beginning in the seventeenth century, with opinions differing as to whether or not it has ended, and if it has when it did so. Besides the rise of science, the beginning of this period is marked by the establishment of a new way of organising society – the economic order known as capitalism, and the individualistic view of human beings is associated with this in particular.

The first attempt to move beyond the world that had taken shape in the seventeenth century was made towards the end of the next century by *Romanticism*. In some ways the ideas promoted by Romanticism anticipated those of the environmental movement today – anticipated and to an extent inspired, of course. That is where I begin. I attempt to say enough about the early Modern world-view to make the Romantic critique intelligible, leaving some of the detail to be returned to in the context of later discussions.

In Chapter 2 therefore I explain what people have said about the nature of science, which is what follows, they say, when the assumption is made that there is a nature independent of human life and interests. And in Chapter 3 I outline that first sustained reaction to the picture of the universe that science developed, the reaction that came at the end of the eighteenth century, and which is known under the unfortunate name of 'Romanticism'.

The Romantics sought to reject what they saw as a whole configuration of ideas characteristic of the eighteenth-century – or 'Enlightenment' – outlook: the presumed inertness of the fundamental constituents of the universe, and of the human mind, the reductionism involved in seeing shape, size and motion as fundamental, its conception of reason and of the place of feeling, and its deterministic cast. But from this comprehensive critique the thinkers today who are closest to the Romantic sensibility draw two major conceptions, both to do with what it is to be a human being, and both rejections of individualism. Their first claim is that the human being is not really to be conceived of on his or her own at all, but as something that in a deep sense belongs to 'nature'; and second, that the point of human life is 'self-development', the development of the self that comprises 'nature' and the human being together.

There is a line of thought there that could provide, if it convinced us of its coherence, a powerful and truly radical alternative to received ways of thinking about ourselves and 'nature'.

A different style of critique has come from writers who have linked the over-exploitation of the environment in the Modern West with its exploitation of other kinds, notably of women. But few contributors from this direction absolve science. Science is there for most of them, involved in one way or another in the multiple exploitations characteristic of Western society. This is the topic of Chapter 4.

If science requires revision, or radical supplementation, or replacement, the framework of thought that might help us, it is argued, is *phenomenology*, to which Jane Howarth, in Chapter 5, offers an introduction.

In the sixth chapter I turn from science to individualism and explain how this underlies the great bulk of Modern Western thinking about ethics and politics.

Chapter 7 and the next explore the resources in science itself to combat the over-exploitation of nature. In this chapter it is the biological conception of life and the evolutionary origins we have in common with living things in general that are the focus – supporting, as some argue, the conclusion that our duty to respect others extends to the boundaries of the living world. This is an argument that works within science, and works within individualism. I then explore (in Chapter 8) the support environmentalists have got from the science of ecology in their plea for a more 'communitarian' approach to nature.

In Chapter 9 I take further the question of what makes, for example, an ecosystem, or an organism, or the earth, *one* thing rather than a mere *collection* of things, and what the significance of this might be. In the pre-scientific framework the one thing/many things distinction was supported by the notion of a *form*. There is a relevant distinction to be made today based on the possession of a unifying *goal*.

A minor disaster perhaps in the scheme of things, where poverty and illness and death are the standard payments, but a disaster nonetheless, is the destruction going on apace of the beauty to be found and enjoyed in nature. The final chapter, contributed by Emily Brady, is devoted to explaining how traditional thinking in aesthetics, which has concentrated on works of art, is being extended to help us think about the importance of natural beauty and so recruited to the cause of thinking out what on earth to do.

My attempt throughout has been to engage the reader in thinking through these large ideas, about which of course, in one guise or another, a great deal has been written. The reader will not emerge with a comprehensive grasp of this literature, nor indeed with a comprehensive guide or introduction to it, but, I hope, with some lines of thought that are open, and promising, and an indication of where to look for help.

The most appropriate entity to thank as this book goes to press is probably the Department of Philosophy at Lancaster, and especially Michael Hammond, Alan Holland, John O'Neill, Helen Shaw and Eileen Martin.

Further reading

Attfield, R. and Belsey, A. (eds), *Philosophy and the Natural Environment*, Cambridge, 1994, CUP.

Botzler, Richard G. and Armstrong, Susan J., *Environmental Ethics*, Boston, MA, 1998, 2nd edn, McGraw-Hill.

Chappell, T.D.J. (ed.), *The Philosophy of the Environment*, Edinburgh, 1997, Edinburgh University Press.

Graves, J. and Reavey, D., *Global Environmental Change*, London, 1996, Longman.

Midgley, Mary, *Animals And Why They Matter*, Harmondsworth, 1983, Penguin.

Rolston, Holmes, III, *Environmental Ethics*, Philadelphia, 1988, Temple University Press.

Westra, Laura and Lemons, J. (eds), *Perspectives on Ecological Integrity*, London, 1995, Kluwer.

2 Objective nature

Many of those who are worried about the present state of 'the environment', and what they see as the catastrophic breakdown round the
corner, blame, in one way or another, *science*. Some think what is
needed is reform. We need to retain science, but also bring about
internal changes. However, there are others who seem to be arguing
for a root and branch elimination.

To evaluate this deep-going and far-reaching proposal you need to
develop a view of what is fundamental to science. This is the point of
the present chapter.

I shall present three ways of thinking about science and its distinctiveness, very different approaches that nevertheless converge on a
single conclusion, which is that at the heart of science is the idea of
an *objective reality* that exists independently of human (or any other)
perception of it, and independently of any webs of meaning and
significance that human beings may weave or cast over it.

The three approaches are the following. They say that science was
made:

1 By the construction of the experiencing subject, relating to the
  'external world' only across a gap.
2 By people beginning to think there was more to the world than a
  compendium of messages and morals for the benefit of human
  beings, and that it might be studied as something 'in its own
  right'.
3 By the peeling off of significance from the world.

The construction of the external world

In one of the most thought-provoking suggestions it is claimed that somehow 'the external world' as we think of it today was not quite recognised in the pre-Modern period. This is thought to be highly significant because, once recognised, 'the external world' is something towards which attitudes can be formed; and a possible attitude is that it doesn't matter very much. So you have created in the conception of the human subject 'looking out' on an 'external world' the pre-condition of an exploitative attitude towards it.[1]

Responsibility for the drawing apart of the human subject, or 'experiencer', and the world experienced is usually attributed to the seventeenth-century thinker, René Descartes.

It is really quite difficult to explain the idea – but not because the subject/external world conception is alien. Explanation is difficult for the very opposite reason: it is difficult to imagine that there might be an alternative! It is alleged though that there has been an alternative way of thinking, and that it is to be found in medieval thought. So I think the only way is to give some account of the theory of *forms* which, as the main structure of the Scholastic intellectual framework, played the key role in their account of human experience. Later, I will have to say some more about the Scholastic form, but here I will say enough to set out the Scholastic account of perception as an indication of the Scholastic approach to human experience in general.

A pre-Cartesian account of seeing

For Aquinas, the part of us that is responsible for perception is our 'intellect', and the problem, as the Scholastic Aquinas conceived of it, was that while the things we see are *material* our intellects are *not*. So, as far as perception was concerned there was a *gap* to be bridged – a gap between material and immaterial.

The notion of 'form' was relevant to the problem Aquinas was addressing, because the form was not something material. The form gave organisation to the material, but it was something beyond the material. It therefore represented common ground between the thing perceived and the intellect: both form and intellect were non-material.

The relationship between perceiver and perceived from the Scholastic point of view is not the relationship between one thing separated spatially from another thing. It is more like the relationship between the intellect and something the intellect has *understood*.

The Cartesian framework

In the Cartesian framework, which was to replace the Scholastic, perception of something was thought of as taking place in virtue of it emitting or reflecting a beam of light, which impacted upon the perceiver having crossed the spatial gap between perceiver and object perceived. Contrast this with how we think of 'understanding'. We do not assume that in order for a sentence to be understood it has to send out 'beams of meaning' for our minds to receive. It is more that somehow we come to share something that the sentence has. The Scholastics thought of perception on this model. Perception was sharing – coming to share in the *form* of the object perceived.

What we have before us is the sense in which Descartes created two worlds where before there had been one. He did not dispense with the Scholastic distinction between material and immaterial, but he added to that a new distinction: the inside world of the mind, to which one person has unique access, and the 'external world'.

I made the point in terms of 'perception'. But the new way of thinking of inner and external worlds separated by a gap played a similarly central role in other kinds of what we today subsume under 'thinking': imagining, planning, meditating, dreaming, and so on.[2]

In the Cartesian revolution, I am tempted to say that the relationship of the human being to the world was re-conceptualised. But that is itself to see things from the new perspective. It is our Modern (that is, after the emergence of modern science) picture of the human being as an entity distinct from 'the world', and on that account constituting the kind of thing that must have some sort of 'relationship' with the world that is the seventeenth-century innovation.

A possible metaphor for the change wrought by Descartes is the following: Think of a patient in bed in hospital. Now think of the nursing staff drawing screens round the bed to provide a degree of privacy. Before the screens get put in place, the patient is 'in touch'

Box 2.1

The modern mind

[A] single inner space in which bodily and perceptual sensations...mathematical truths, moral rules, the idea of God, moods
of depression, and all the rest of what we now call 'mental' were
objects of quasi-observation.

(Richard Rorty, *Philosophy and the Mirror of Nature*,
UK edn, Oxford, 1980, Blackwell, p. 50)

with the rest of the ward in a way that stops being true afterwards.
Afterwards, the patient sees nothing of the rest of the ward – except
for whatever shadows appear on the screens. You could say that the
new thinkers drew screens around the human being. From that point
on the shadows cast on the screens by objects beyond had to take the
place of the direct communion with the ordinary things that had
been assumed before.

This is to make the point in terms of just the one function – perception. However, it applied to the new way of thinking about human
experience generally. Two worlds were sharply distinguished. As an
experiencer, I am directly in touch with the contents of my mind, that
is, one world, the inner world, at the centre of which I live. But
beyond my mind is the *external* world. Different people had different
ideas as to how the two worlds related. Some thought you could
really have very little knowledge of the external world – or indeed
none at all – while others thought that you could rely (for reasons
they put forward) on the contents of your mind being an accurate
representation of things in the external world, so that the external
world was to some degree accessible to you.

This Cartesian picture is often the direct target of environmentalist
criticism. It is a picture of the person, or at least the mind in the
person (sometimes called the Cartesian 'subject'), as utterly distinct
from the physical world. In so far as any dealings between the
'subject' and the 'external world' are thought by the Cartesian to be
possible, they are pictured as going on via intermediaries (Descartes'

Box 2.2

Representation

The notion of a representation is so crucial to understanding the Cartesian shift that I need to make sure it is clear. There is some danger that its very familiarity may get in the way of seeing its significance.

What is meant by representation?

Think of what we learn about dreams, for example. Children start off thinking that what we call dream experiences are for real, don't they? They believe, from what they say, and from how they behave, when they wake up, that the monster was actually in the room. What do adults actually say to them? That the experience of seeing the monster was not a contact with the real world, or with any world outside the dreamer, but the dreamer dealing with figments. What passed before the dreamer's mind were not real monsters, but images of monsters, non-substantial stand-ins for monsters, or representations of monsters.

Some cultures conceive of dreaming differently, don't they? A dream is sometimes regarded as giving access to reality. The dreamer is a specially gifted person, a shaman, a person gifted with insight: one who is in touch with a different world.

I'm not saying who is right. I'm just trying to give you a picture of what I mean by 'representation'. We think of dreams involving images, mental pictures passing before the dreaming mind. The cat we dream of is not a real cat. The real cat is in the chair downstairs. We just have before our dreaming mind an image of the real cat, a representation of the real cat, some kind of stand-in for the real cat.

When we are trying to solve a problem in our heads, we presumably deal in stand-ins too. If we are thinking say of how to get from here to the tower, we imagine ourselves going down this road and that, turning left, turning right, and so on. These roads that we think of ourselves going down are surely not real roads, but they are out there somewhere. What we have in our heads are stand-ins for roads, a mental map perhaps, some kind of representation or model of the road system in our heads.

The modern conception of an idea, from Locke building on Descartes, is of mental items as representations. Thus, we are directly in touch as we 'look out' onto the 'external world' not with things themselves, but with their stand-ins or ideas. And the agenda for modern epistemology is set: How can we build knowledge of 'the external world' if that is the limit of our contact with it?

'ideas'). This is a framework of thought, say the critics, in which exploitative attitudes towards everything on the other side of the great divide will flourish. The scene is set, by the Cartesian construction of the mind, for 'the environment' to be regarded as fundamentally alien. And what is alien has no hold on us; no appeal to our concern.

The modern world as 'objective'

The same point is arrived at from a very different direction by those who describe the key feature of the emergence of the 'modern' outlook as a sense of the *objectivity* of nature. They argue that, for the pre-modern world, nature was always thought of in relation to human beings, whereas from the seventeenth century onwards an interest developed in nature as it was independently of the 'meaning' it carried for human beings. The growth of science was the growth of knowledge about animals and plants and the rest of nature as they were 'independently'. Let me explain by drawing on the work of the historian Keith Thomas, whose *Man and the Natural World* [3] presents in engaging detail some of the features of the modern perspective on animals and plants, and the shifts in that perspective that have taken place from the seventeenth century.

Before the rise of modern science, Thomas argues, knowledge had been organised around the practical problems human beings faced. For example, the interest in plants was largely rooted in an interest in their medical uses, and the way in which they were classified reflects this. For example, in 1526 the *Grete Herbal* divided mushrooms simply into two types: 'one…is deadly and slayeth them that eateth them;…and the other doth not'. [4]

The situation with animals is parallel. Topsell, for example, in his *Historie of Foure Footed Beasts* of 1607 explained that his main point was to show which beasts were the friends of human beings, which could be trusted and which could be eaten. Another example is given in Box 2.3. [5]

Box 2.3

A classification of dogs

Generous

Used in hunting

- good at smelling
- good at spying
- good at speed
- good at subtlety
- good against beasts
- good against birds
    - good against water birds
    - good against land birds

Used by fine ladies

Rustic

Sheepdogs
Watchdogs

- those that bark
- those that bite
- those that bark before they bite
- those that bite before they bark

Degenerate

(Drawn from John Caius, *Of English Dogges*, UK edn, 1576)

What developed with the modern revolution, Thomas claims, was a challenge to this anthropocentric point of view: maybe (the thought was) things had a structure, a working one, that was interesting for its own sake. The more distant aim of the new science, as articulated by writers such as Francis Bacon, might have been to put it to work in making life easier. But the immediate object, and the way of achieving the ultimate goal, was to gain an understanding of animals, plants, and rocks, etc., based on the premise that there were things to understand that were quite independent of human needs and practical concerns.

Thomas identifies this as the fundamental shift that marked the emergence of modern science, and of what we call the 'modern' world. At the end of the medieval period it was conventional to regard the world 'as made for man and all other species as subordinate to his wishes'. What emerged was 'the idea that nature had a structure independent of human beings, and one that should be studied'.[6]

At first sight this account of what is essential to science seems to wrong-foot a familiar environmentalist line of argument. It is often argued that our environmental troubles spring from an unduly *anthropocentric* outlook on nature. The problem is, it is often said, that we always look on everything from the human point of view, as though everything were there just for us. But if Thomas' account of the nature of the scientific revolution is to be believed, it was exactly anthropocentrism that was jettisoned by the new science. What the revolutionaries rejected was the idea that everything in nature was the way it was either to provide for human beings' bodily needs, or in order to teach us moral and religious truths. This would seem to imply that if exploitative attitudes are rooted in an anthropocentric outlook, science cannot be blamed for them.

Foucault's account of the constitutive feature of the modern world

The third attempt at identifying the defining feature of the seventeenth-century revolution is the one developed by Michel Foucault. His idea is that the modern world began with a split between *language* and *the world*.

Foucault looks at the accounts of animals and plants from the medieval world, and sees that what is presented is 'a unitary fabric'. Woven together, he says, were 'all that was visible of things' and also '*the signs* that had been discovered or lodged in them'.[7] Signs were included in pre-modern accounts according to Foucault because *they were regarded as parts of things themselves*. In the seventeenth century signs become 'modes of representation'. This was the revolution, according to Foucault.

An example of the way in which all sorts of things (to modern eyes) were included in pre-modern natural history (that is its 'heterogeneity') is provided by Francis Bacon. Bacon includes 130 topics for inclusion in his *Natural History*, of which perhaps two dozen fall within the aegis of natural history as it came later to be understood. He lists for example:

- History of the manufactures of Silk, and the arts thereto belonging.
- History of manufactures of feathers.
- History of Dyeing.
- History of Waggons, Chariots, Litters etc.
- History of hunting and fowling.
- History of Jugglers and Mountebanks.[8]

Natural history could encompass even parable, as Plate 2.1 shows. According to the *Bestiary* in which this picture appears, the ostrich waits each year for the appearance of a particular constellation of heavenly bodies:

When, in about the month of June, it sees those stars, it digs in the earth, lays its eggs and covers them in sand. When it gets up from that place, it at once forgets them and never returns to its eggs...If the ostrich thus knows its proper time, and forgets its offspring, laying aside earthly things to follow the course of heaven, how much more, O man, should you turn to the prize of the summons from on high, for which God was made man.

Plate 2.1 *The medieval outlook often saw the facts of natural history as parables*

Source: Picture of the ostrich, illustrating its reputed abandonment of its eggs, from *Bestiary*, MS Bodley 764, fol. 67r; as appearing on p. 73 of Lorraine Daston and Katherine Park, *Wonders and the Order of Nature, 1150–1750*, Zone Books, 1998, New York

Another example is provided by Aldrovandi's (1522–1605) treatment of the serpent (see Box 2.4).

This enormous 'heterogeneity', as it appears to modern eyes, seems quaint. These eyes of ours see

Box 2.4

Aldrovandi's natural history of the serpent

This has the following subheads (from U. Aldrovandi, *Monstrorum Historia*, 1647, cited by Foucault in *The Order of Things*, p. 39):

- equivocation (various meanings of the word 'serpent')
- synonyms and etymologies
- differences
- form and description
- anatomy
- nature and habits
- temperament
- coitus and generation
- voice
- movements
- places
- diet
- physiognomy
- antipathy
- sympathy
- modes of capture
- death and wounds caused by the serpent
- modes and signs of poisoning remedies
- epithets
- denominations
- prodigies and presages
- monsters
- mythology
- gods to which it is dedicated
- fables allegories and mysteries
- hieroglyphics emblems and symbols
- proverbs
- coinage
- miracles
- riddles
- devices
- heraldic signs
- historical facts
- dreams
- simulacra and statues
- use in human diet
- use in medicine
- miscellaneous uses

natural histories in the old style as little more than rag-bags. Buffon in the eighteenth century expresses the modern reaction when he complains that natural histories of the early period are stuffed with a 'vast amount of useless erudition, such that the subject which they treat is drowned in an ocean of foreign matter'.[9]

Foucault turns this observation on its head. The editors of these compilations did not regard the material as heterogeneous, he points out: they thought it all belonged together. Distinctions that came to be important later, for example between observation, document, fable, had yet to be drawn.

Plate 2.2 *Another medieval attitude was to see nature as symbolising theological truths*

Source: Picture of the panther, illustrating its reputed sweetness of breath, from *Bestiary*, MS Bodley 764, fol. 7v; as appearing on p. 42 of Lorraine Daston and Katherine Park, *Wonders and the Order of Nature, 1150–1750*, Zone Books, 1998, New York

Religious symbolism was not seen as distinct (see Plate 2.2). The breath of the panther, we are told, is so attractive that 'when the other animals hear his voice they gather from far and near, and follow him wherever he goes…Thus our Lord Jesus Christ, the true panther, descended from heaven and saved us from the power of the devil.'

For Aldrovandi, says Foucault, 'nature, in itself, is an unbroken tissue of words and signs, of accounts and characters, of discourse and forms'. For him, to write an animal's history:

> [O]ne has to collect together in one and the same form of knowledge all that has been seen and *heard*, all that has been *recounted*, either by nature or by men, by the language of the world, by tradition, or by the poets…Aldrovandi was neither a better nor a worse observer than Buffon; he was neither more credulous than he, nor less attached to the faithfulness of the observing eye or to the rationality of things. His observation was simply not linked to things in accordance with the same system or by the same arrangement of the episteme. For Aldrovandi was meticulously contemplating a nature which was, from top to bottom, written.[10]

So Foucault's thesis here is that words (a sort of sign) are not separate from nature in the pre-Modern period, but intrinsic to it, woven in with everything else to make a single cloth, and it is language's splitting off from the world that constitutes this most seminal of shifts between pre-Modern and Modern thought structures.

Prior to the Modern framework, language and discourse are part of nature residing 'among the plants, the herbs, the stones, and the animals'.[11] Under the Modern framework, language is an *independent* system of signs which can be used to *represent* nature. That is, the splitting off of language from the world creates a tool whereby items in the world can be represented. And once you have representations of items in nature, those representations can be manipulated and organised in different ways. Thus, the splitting off of language from nature creates the possibility of ordering it. Before, words were part of the world. Afterwards, they were the tools by which things in the world and their relations were represented.

This ordering of things, which language splitting from the world makes possible, is, according to Foucault, the great project of the first period of Modernity.

Box 2.5

The new science

[T]he fundamental element of the classical *episteme* is neither the success or failure of mechanism, nor the right to mathematicize or the impossibility of mathematicizing nature, but rather a link with the mathesis which, until the end of the eighteenth century, remains constant and unaltered.

(Foucault, *The Order of Things*, Tavistock edn, p. 57)

What Foucault appears to be saying is that with the seventeenth century a new general science of order was created, and that this was thought to embrace all possible knowledge.

Implications of this perspective for our dealings with (what had become) 'the environment'

These accounts of the fundamental features of the framework of thinking brought in by the scientific revolution in the West in the seventeenth century, all in their different ways, give a central role to 'objectivity'. The Cartesian construction of 'the mind' constructs at the same time an 'external world' independent of human experience. Thomas observes a new interest in things as they are independently of human beings. Foucault, with more drama, argues that the revolution sees the splitting of one world into two: one bereft of meaning, the other a world of signs – 'analytical tools' which human beings can use to represent things in the other world, and to construct different 'orderings' of them. These three points are enormously different in tone, but they support the same idea – that at the heart of the revolution was a recognition, or a construction, of a world of nature possessing an existence apart from human beings and their understandings.

Many commentators have claimed that we need look no further for the origins of the modern Western exploitative attitude towards nature. Without this objectified conception of 'nature', our world would be 'part of us', and damaging it would be damaging ourselves. As something *alienated* from us, set *apart* from us, nature is something we don't intuitively care about. Even if it had an 'interest' of its own, the interest of an alienated nature would not be *our* interest, and would be one we might easily ignore. *This*, it is said, is the fateful conceptual shift that sets the agenda for environmental philosophy, which is so much preoccupied (for good reason) with the question of whether anything other than human beings has 'value'.

But there have of course been other diagnoses, staying with the idea that it is science that is at the heart of the problem, but offering different understandings of what it is about science exactly that encourages damaging exploitation.

Special significance has been attached, for example, to the way in which scientific thought (allegedly) represents the universe as a piece of *mechanism*. Nature, according to science, it is argued, is a machine, whereas under previous outlooks it had been regarded as more like a living thing or organism. It is suggested that machines invite exploitation in a way that living things do not.[12]

After a first tremendous run, all aspects of Modern science came under a sustained barrage of criticism towards the end of the eighteenth century. Some of the most radical thoughts of environmentalists today are bringing the ideas of the revolutionaries of that period back into play. I think it would be helpful to look at them in the hope that the distance between then and now will lend perspective and clarity.

Questions

1 Do you see seeds of our environmental situation in the scientific revolution of the seventeenth century?
2 Could the universe have a meaning even in the absence of a God?

Further reading

Foucault, Michel, *The Order of Things*, UK edn, London, 1970, Tavistock.

Merchant, Carolyn, *The Death of Nature*, London, 1980, Wildwood House.

Rorty, Richard, *Philosophy and the Mirror of Nature*, UK edn, Oxford, 1980, Blackwell.

Thomas, Keith, *Man and the Natural World*, London, 1983, Allen Lane.

3 We are all one life

The large idea I want to build towards in this chapter is that in the modern West we are fundamentally wrong about ourselves and our place in nature. Somehow, it is said, human beings have set themselves apart from nature, and it is this that leads to the dangerous ways we have of exploiting the world about us.

Understood properly, human beings are *part* of nature. If we understood that, we would understand that destroying the prairie or exterminating the wolf or polluting the sea are all forms of *self-mutilation*. In so far as we are part of nature, our well-being is an aspect of the well-being of nature as a whole. John Donne's famous lines refer most obviously to the community of human beings to which he is urging we should all remember we belong. But – it is said – there is a wider point to be made. As human beings, we are parts not only of the community of humanity, but also of the community which makes up nature as a whole. 'We are all One Life', in the words of Coleridge.[1] So the bell tolls for us not only when a fellow human being dies, but at the destruction of any member of that vastly wider community, which is nature itself.

The line of thought here, which you may have encountered, covers a range of views not all of which are really radical. At one pole, the idea that human beings are part of nature is little more than a reminder that 'he who shits on the path will find flies on his return', as the Yoruba proverb has it. In one sense, such a reminder might be the most important thing to reiterate and amplify: Be careful! Think of the medium and long-term consequences of what you do! It only needs for us human beings to adjust our behaviour (including the behaviour of our states and corporations) in the light of that admonition, and the human-made threat to our survival would be removed.

At the other pole, the unity that is proposed to embrace in one nature human beings on the one hand and all the rich variety of things that are usually located in the 'environment' on the other involves a decisive break with orthodox ideas.

The most sustained development of the idea that 'we are all one life' comes in contemporary philosophy from a school of thought that calls itself 'Deep Ecology', which is associated first and foremost with the name of Arne Naess. This school adds to the idea of unity the notion of 'self-expression' as a goal: the unity that is nature is motivated by a drive towards self-expression.

These concepts are not new, however, and it will, I think, be helpful to approach them through an earlier movement of thought, namely the discontent, voiced towards the close of the eighteenth century, with what had become of 'science' – the 'Romantic' movement. 'Romantic' is an awkward name for the set of ideas here, but we appear to be stuck with it.

The Romantic critique of Enlightenment science

What you have in the Romantic movement is an attempt to reject the conception of the world and of the human place in it that had been sponsored (as the Romantics saw it) by the pioneers of Modern science (such as Bacon and Descartes) in the seventeenth century, and carried through as the 'Enlightenment' into the eighteenth century. Part of the benefit to us of exploring Romantic thought is that it gives us the occasion to review the mind-set it was reacting against – the mind-set of the Enlightenment. In spite of the best efforts of the Romantics, large elements of the Enlightenment conception are part of our outlook today, and some of the critics of contemporary thought, the proponents of Deep Ecology for example, may be regarded as renewing the Romantic critique.

The perspective of the Enlightenment was to regard Modern science in its earliest days as having generated an unprecedented increase in knowledge in the fields of mathematics, astronomy and physics, and as having the potential of generating the same kind of revolutionary progress as it was applied elsewhere. The Enlightenment thinkers saw their age as one in which Modern science, having been launched in spectacular fashion by the pioneers, was now advancing towards maturity – advancing through its application to *every* field where knowledge for human beings was held to be possible. The general attempt to apply science throughout the entire domain of human knowledge is sometimes called the Enlightenment project.

Towards the end of the eighteenth century, a reaction set in. It took the form not of the rejection of science, but of the demand that science should reform. It is not clear that it got very far. It sponsored new concepts that became influential, and science certainly took a turn as the century turned. (Foucault calls the turn a revolution indeed; one that rivalled the birth of Modern science in significance.) What is unclear is whether the new direction was one that the Romantics would have welcomed.

As they looked back, anyway, those looking for a revolution towards the close of the eighteenth century saw six features of the familiar science which they wished to reject.

Human beings and nature in Enlightenment thought

The universe and its constituents as inert

First, established science was seen as assuming that the universe and its basic elements were *inert*. The distinction between *inert* (or *passive*) and *active* was clearly recognised in medieval thought. A passive body was one that was simply washed about in the tide of events. It might move because it was pushed, or perhaps because it was pulled, but, insofar as it was indeed 'passive', it 'took no initiative' of its own, as it were. An active thing, on the other hand, was one that was capable of *launching* into movement spontaneously. Active things were thought of as possessing a 'power' inside them that was capable of initiating change. In the Enlightenment period it became possible to doubt that there were such things.

Thomas Hobbes (1588–1679), for example, held that everything that happened in the universe could be explained in terms of masses of small particles which were perfectly passive: they moved about, but only because they kept being bumped into by others.

This is 'atomism' in a particularly clear form: the universe is made up entirely of indivisibly small particles – jostling about, but perfectly inert – and everything about the universe, all the changes that go on, as well as all the appearances it presents to us – of colours, smells, sounds, and so on – are to be explained in terms of these minute blobs and their senseless jiggling.

There *were* Enlightenment thinkers who were 'atomists', but who believed the atoms were *active* (Leibniz, at one point in his career at any rate, was one of these). Nevertheless, the passive conception predominated, and it was this that entered into later conceptions of how the universe was thought of by the Enlightenment: it was thought of as made up of minute, hard, *passive* particles.

The predominant form of Enlightenment atomism gives us one clear example of what it was to think of nature as in itself 'inert'. There is another great picture of passivity rooted in the Enlightenment, but meaningful to us still today, and that is the metaphor of the universe as a great *machine*. This too invites the idea that nature is passive. Clockwork is a contrivance of parts, each of which is brought into movement by something else. None of the parts 'takes any initiative' of its own.

The human mind as inert

Part of 'the Enlightenment project' was to bring to bear the scientific mind-set upon the human being itself, and when this was done, generally speaking, the passive conception of nature was taken. One influential approach was to think of 'ideas' as passive, mental 'atoms'. Sensations dropped through the letterbox of the senses, then through various mechanisms linked up in various ways. The object of the science of the human understanding was to work out the principles which governed the way in which such data of sense combined with and otherwise related to each other – just as it was the object of the science of the natural world to establish the principles according to which atoms (or *corpuscles* as they were called) interacted to produce the observed phenomena of physics.

The universe reduced to shape, size and motion

A third aspect of the Romantic conception of the 'Enlightenment' mind-set was the following. A part of atomism was – and still is – the idea that much of the ordinary experience we have as human beings is somehow a product of the interaction of atoms which, in themselves, have only a small number of simple properties. Atomism proposes to explain our experience of *colour*, for example, by reference to the movement of atoms which themselves are not coloured at all. Sound is put forward as some sort of movement of atoms. Smell is similarly put forward. Different atomists give different lists of the properties that atoms are supposed to possess – and sometimes the same atomist gives different lists on different occasions! But the general thrust is to *reduce*. The world of human experience turns out, according to atomism, to be much more varied and rich than the reality underlying it all, which is, for most atomists, a colourless, silent, odourless, tasteless skitter of tiny specks.

'Reason' in the Age of Reason

Also, in its reaction against the outlook of the Enlightenment, the Romantic revolution focused on the role it had given to 'reason'. The Enlightenment is often known as the 'Age of Reason'. One justification for this name was that the period had turned its back on the authority of the Church and the Bible, and sought to base knowledge instead on whatever could be discovered and verified by human beings on their own. 'The motto of the Enlightenment', Kant tells us, is this: 'Have courage to use your *own* understanding.'[2] In this context, 'reason' stood for the power of unaided human thought. Part of Romanticism's re-evaluation of 'reason' had to do with this. We had better in some contexts, they taught, rely on things outside ourselves as a source of authority for our beliefs (on 'nature', for example, as I shall explain).

But 'reason' is sometimes used to make a different contrast, not between finding out ourselves and taking things on authority from something or somebody else, but between the analytical aspect of human thought and 'emotion'. The Romantics thought they found an emphasis on reason in this sense in the Enlightenment too. They saw it, for example, in the key role that had been ascribed, particularly in early phases of the scientific revolution, to mathematics. Some of the pioneers had certainly seen in mathematics the one and only key to understanding, so that science for them amounted to the progressive application of this form of reasoning to all the possible fields of human knowledge.

But, though some early Modern thinkers thought the whole secret lay in mathematics, as science developed throughout the eighteenth century, it did so by means other than the progressive application of mathematics. It was Natural History that in the eighteenth century displaced physics as the leading science, and Natural History had no mathematical dimension. Its project, under the leadership of Linnaeus and Buffon, was to lay out the order of nature. They did this not by discovering formulae, but by creating elaborate classificatory schemes in which every form of life could be ascribed a particular place.[3]

The Enlightenment placement of feeling

Though non-mathematical, such a project as Natural History was rightly identified by the Romantics as relying on 'reason' as contrasted with feeling. Classifications appealed to properties that could be observed through the senses: especially shape, size, how the parts of a whole were arranged in space and how many there were. Plants were never classified on the basis, for example, of how they made you *feel* – of the pleasure they prompted as you encountered them, or of the emotions they conjured up within you as they were eaten. That whole dimension of the role of plants in our lives was discounted as not relevant to the project of science.

Human emotion *did* enter into Enlightenment discourse in one form however. It was thought to play a part, and a key part, in the determination of human action. David Hume (1711–76) is famous for his view that it is emotion and not reason that provides the motivation for action, and many other thinkers of the period thought of the relationship between human emotion and action in the same way. It is when we ask about what determined the emotions in their turn that we see what the Romantics were objecting to. For Hume and other Enlightenment thinkers saw emotion in quasi-mechanical terms: We were born with a repertoire of emotional responses, and particular responses were triggered by eventualities impinging upon us. Once triggered, a response then motivated action.

This treats the emotions as cogs in a machine. They play a part in the machinery which produces action, but they don't *originate* it.

Determinism in Enlightenment thought

In thinking of the human being as a 'subject for science', Enlightenment thinkers seemed to be committed to the idea that what needed to be understood about human behaviour was its *causation*. I have said that a leading idea was to treat thought in the same way – what was to be understood was what brought a particular thought about, and to regard thoughts, including feelings, as playing a part in the machinery which brought about behaviour. But this way of thinking of the human being brought with it the suggestion – the implication perhaps – that human beings were at the mercy of causal forces. In this way, human *freedom* seemed to be under threat. If what you did was a result of the sort of causes that it was the business of science to identify, this seemed to take responsibility for your actions out of your hands. It wasn't just feelings that seemed like cogs in the clockwork from this point of view – it seemed implied that the human being as a whole was a component in some gigantic machine.

The characteristic of a machine is that each part of it comes into play when caused to do so by some other part acting upon it. Without a prompt, a part remains motionless. So in a machine, nothing happens without a cause. This nostrum, when applied to the universe as a whole, is the thesis of determinism. Determinism's thesis that 'every event has a cause' means that the occurrence of any event can be tied to a change in the circumstances that obtained at the time of its occurrence. In Leibniz's words: 'Everything remains in the state in which it is if there is nothing to change it.'[4]

Determinism stands against the idea that some things that happen, happen, as it were, *spontaneously*. Their occurrence is not fixed by prior conditions. It has seemed to some that human freedom depends on there being *some* of these 'spontaneous' events – namely human 'acts of will'. Human acts of will, in this view, aren't determined by what has gone before. They are 'free'. And only if we think of them as free, can we think of human action as anything other than bits of the universal machine turning.

The power to *initiate* change, in this view, exercised by human beings when they act freely, means that sometimes a change happens while the circumstances in which it occurs remain unchanged. It is clearly a peculiar power: It is a power to bring something about, but one that is thought of as somehow not belonging to the circumstances within which the event in question occurs. It is nevertheless this power that the Romantics wished to have recognised, and for the neglect or denial of which they castigated science.

Human beings and nature in 'Romantic' thought

The Romantics sought to insist then on a conception which made the human being the *originator* of activity, and not simply a node in a causal nexus. 'Everyone's actions', says Herder, 'should arise utterly from himself.'[5] If human beings have no power to initiate change, they are mere 'playthings' of forces impinging upon them, and to be a plaything, the early *Sturm und Drang*[6] writer, Lenz, put it, 'is a dismal, oppressive thought', amounting to 'an eternal slavery, an artificial…wretched brutishness'. Instead, we should place the capacity to act at the centre of our conception of the human being: 'action, action, is the soul of the world, not enjoyment, not sentimentality, not ratiocination'.[7]

Romanticism rejects the notion that the human mind is passive

A part of this new insistence on the power of human beings to initiate change amounted to a new view of the nature of the human *mind*. The mind had been a billiard table on which ideas cannoned about. The revolutionary view was that the mind took initiatives.

The poet Coleridge (1772–1834) – the Coleridge we know as a poet, anyway, though he also thought of himself as a philosopher – articulated his distinction between *fancy* and *imagination* as a way of making the point. The fancy is little more than a spectator in the theatre of the mind, whereas the imagination, as Coleridge conceives

of it, is 'a living Power', creating new forms. Discrete ideas, simply
observed by the fancy, or linked loosely together in temporary assemblies, are by the imagination, as Coleridge understands it, melted and
cast anew. He asserted that the imagination was *vital* – meaning that
it had that power to initiate change often associated with things that
were alive.

The creative power that is the imagination is central to the human
mind, says Coleridge. It is the 'prime Agent of all human Perception',
and thus he understood the mind as essentially creative, in contrast to
the picture of the mind he found in Newtonian science, 'a lazy
Looker-on on an external World'.[8]

Kant, from whom Coleridge derived much of his (philosophical)
inspiration, broke with Enlightenment thought on this point too –
the mind was no mere spectator – it played a part in creating the
world as we know it.

Romanticism thinks of growth as self-realisation

Romantic thinkers had an alternative to put in place of the conception of the human being as a cloud of jostling particles. They
proposed instead the metaphor of the seed and its development into
a mature plant.

This metaphor encapsulated a quite radical innovation. It gave a way
of thinking of the human being as in origin inchoate, or lacking
form, but as then gradually acquiring structure and differentiation as
development towards maturity proceeded. This understanding was
applicable to physical development, but the Romantics also applied it
to the development of the person as a whole. They recommended
understanding the *person* as something that began unformed, and as
something that *acquired* form as time passed.

This way of thinking of development relied on the idea of a *potential*
becoming *realised*. In the beginning, the fertilised egg was small and
undifferentiated. Eventually it grew into something with elaborate
differentiation and organisation. Development towards maturity
could be regarded as the gradual realisation of that original potential. This was how the Romantics also saw the life of the person as a

whole – a movement towards the realisation of potential. They saw it as a process of *self-realisation*, a process that begins with a real but inchoate self, and proceeds through the gradual crystallisation of characteristics and personality which had been 'pointed to' – but *only* 'pointed to' – in the beginning. Self-realisation, for the Romantics, was the point of life.

The new 'Romantic' conception of the self, taking shape here, is thus of a potential that undergoes development. Already, I have pointed out the importance that the Romantics placed on the power of the human being to initiate change. The new self is also characteristically *self-powered*. The drive to development comes from *within*. In the mind, Herder says, 'there are glow forces, living sparks'.[9] It is these inherent, autonomous, self-energising 'forces' pushing their way against the world outside that propel the self into realisation. As Charles Taylor puts it, for Herder, realising the human self involves 'an inner force imposing itself on external reality'.[10]

Therefore, the Romantic self in its primal state is a kind of *seed* that was thought of as destined for development – a process that it would launch under its own initiative.

Romanticism rejects the view that nature is passive

The Romantic movement challenged in this way the idea that the human mind was passive. But nature too, for the Enlightenment, had been passive, and furnished with entities that were passive. Romanticism also challenged this.

The concept it reached for in asserting its alternative view was the concept of *life*. Living things have always suggested the idea of spontaneity. Animals of course have no appearance of being passive. They were characterised by the Greeks as things possessing the power of movement. They might sit quietly for a period, but then they spring into motion, quite often apparently without stimulation. But plants have something of the same apparent power too. They don't generally move about, but they do spring up out of seeds in a way that has always been recognised as distinctive and highly significant. So the Romantics expressed their opposition to the prevailing deterministic view of the natural world by insisting that it was *alive*.[11]

By implication, the insistence that things in nature and everything belonging to it were 'alive' militated against the reductionist nostrums of established science. There were also plenty of explicit statements to this effect. Goethe (1749–1832), who is not classed by historians as a Romantic, but did belong to the late eighteenth century and was a powerful critic of Enlightenment science, conducted an investigation into the nature of colour, and the human experience of colour, because he thought the reductionist treatment offered by Newton was profoundly misconceived. More well known are the pronouncements of the later Romantic poets, such as Wordsworth, with a vision of nature as anything but the endless jostling of colourless, odourless, tasteless, mindless particles.

> Ye Presences of Nature, in the sky
> Or on the earth! Ye Visions of the hills!
> And Souls of lonely places!
> (*The Prelude* 1, 464–6)

Nature is a cathedral of 'Presences' for Wordsworth, not a matrix of particles, as this famous invocation makes clear.

Plate 3.1 *The Romantic attribution of mentality to nature*

Turner: *Ullswater Lake from Gowbarrow Park, Cumberland* (1815). © Whitworth Art Gallery

Romanticism gives a new authority to feeling

Wordsworth highlights the new significance attached by Romanticism to *feeling*. The 'Presences' he has registered in the extract on p. 31 invest, he says, the whole natural world with feeling. He says that we should understand the workings of the 'Presences' in the way in which the wind imparts a swell to the ocean. They make:

> the surface of the universal earth
> With triumph, and delight, and hope, and fear,
> Work like a sea.
>
> (*The Prelude* 1, 490 4, 499 501)

Nature and natural things were in this way thought of by the Romantics (or at least by the Wordsworthians) as capable of supporting 'feeling', but also a new perspective was taken towards the feelings that went on in human beings. As I have pointed out, feeling had not been *neglected* by Enlightenment science – it had been regarded as an important part of the human machine. But for the Romantics it had an altogether different significance. They looked upon feeling rather as a much earlier tradition had looked upon reason – as the guide to right behaviour. 'The heart', as the poet and writer Novalis puts it, 'is the key to the world and life.'[12]

There was a long tradition before the eighteenth century and before modern science which thought of the experiences we now identify as emotions of one kind or another as distracting or misleading. Plato suggests the image of the soul of a person split into parts, with reasoning often in conflict with desiring. Our reason might tell us to do one thing, while our desires tell us not to. For example, our reason might tell us the water is poisoned, but our desire to drink might drive us to swallow the poison nevertheless.[13]

In medieval thought there is the idea that it was the role of reason to tell you what you ought to do, and that a weak person, even knowing what s/he ought to do, might be preyed upon by anger, fear, envy, etc., and stopped from acting as s/he knew s/he ought. Medieval thought conceived of the reason as the compass of a person's ship: It showed the way from which the storms of 'feeling' might mislead.

Therefore, in re-evaluating the role of feeling in human life the Romantics had more than Enlightenment rationality to oppose. The condemnation of feeling went back a long way, and had taken a variety of different forms.

Feelings were important for the Romantics partly because they thought of them as the way in which nature manifests itself to us. Therefore, in heeding feelings people heed the promptings of nature.[14]

Identification of the human being with nature

There is finally in Romantic thought a powerful strain of 'nature-mysticism'.

It belongs to a long tradition of 'mystical' thought in the West, a tradition that lends itself particularly to the strain in Christianity which despises this world and tries to shift the focus elsewhere. An analysis within the Christian tradition describes the mystical experience as one that involves a feeling of awe, of being utterly

---

Box 3.1

The mystical experience

1 The unitary consciousness from which all the multiplicity of sensuous or conceptual or other empirical content has been excluded, so that there remains only a void and empty unity. This is the one basic, essential, nuclear characteristic from which most of the others inevitably follow.
2 Being non-spatial and non-temporal, which of course follow from the nuclear characteristic of 1.
3 A sense of objectivity or reality.
4 Feelings of blessedness, joy, peace, happiness, etc.
5 A feeling that what is apprehended is holy, sacred or divine.
6 Paradoxicality.
7 Allegations by mystics of ineffability.

(From W.T. Stace, *Mysticism and Philosophy*, London, 1961, Macmillan, pp. 110–11)

overpowered, of energy or urgency, of 'stupor' and, finally, a feeling of 'fascination'.[15] However, other traditions appear to recognise much the same kind of phenomenon. The characteristics that are claimed to be common ground for a number of religious traditions have been listed in Box 3.1.

The nature-mysticism, which we find in Romantic writers like Wordsworth, is not generally regarded as on all fours with the mysticism to be found in the world religions. But it shares with them – and this is our interest – something of the expansion or loss of personal identity that they regard as fundamentally characteristic of the mystical experience.

Here is a version of the religious perspective, for example:

As a lump of salt thrown into water melts away…even so, O Maitreyi, the individual soul, dissolved, is the Eternal – pure consciousness, infinite, transcendent. Individuality arises by the identification of the Self, through ignorance, with the elements; and with the disappearance of consciousness of the many, in divine illumination, it disappears.[16]

Plate 3.2 *A Romantic vision of nature*

Blake, *Europe*, plate 1: Frontispiece, 'The Ancient of Days' (1824). © Whitworth Art Gallery

Box 3.2

Mysticism in Plotinus

> No doubt we should not speak of seeing but, instead of seen and seer, speak boldly of a simple unity. For in this seeing we neither distinguish nor are there two. The man…is merged with the Supreme, one with it. Only in separation is there duality. This is why the vision baffles telling; for how can a man bring back tidings of the Supreme as detached when he has seen it as one with himself…Beholder was one with beheld…he is become the unity, having no diversity either in relation to himself or anything else…reason is in abeyance and intellection, and even the very self, caught away, God-possessed, in perfect stillness, all the being calmed.
>
> (Plotinus, *Works*, trans. Stephen MacKenna, New York, New York Medici Society, The Sixth Ennead, Ninth Tractate, §10. Quoted by Stace, *Mysticism and Philosophy*, London, 1961, Macmillan, p. 104)

And, more tersely, from the Christian tradition, in a thought that encompasses the whole of nature as well as the individual soul, Meister Eckhart:

> All that a man has here externally in multiplicity is intrinsically One. Here all blades of grass, wood, and stone, all things are One. This is the deepest depth.[17]

The loss of personal identity, or, put another way, the identification of the self with nature as a whole, explicit in this mot from Eckhart, is the thought that finds particular expression in nature-mysticism. The religious mystic comes to understand that they and God are one; the nature-mystic that they and nature are one. The identity of the nature-mystic expands as it were into nature, so that what is asserted is that the distinction between the individual and nature is lost. The two become one (or what is discovered is that what appeared to be two has really been one all along).

Identification of the human being with nature – Deep Ecology

There are strong echoes of the Romantic critique of the established eighteenth-century outlook in environmentalist thought today, and I shall be explaining some of them in what follows.

One of the most radical modern critiques comes from the 'Deep Ecology' movement, which is founded on two of Romanticism's leading notions – the concept of 'self-expression' and its importance in understanding human life, and the idea that the human being is to be seen as *a part of*, and not *apart from*, the wider whole that is nature at large.

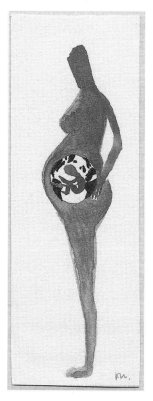

Plate 3.3 Deep Ecology suggests human beings ought to be identified with nature, and the object of the new entity should be self-realisation

Source: Graphic by Mari Fuji

When Arne Naess (b. 1912), as the leading spirit of the Deep Ecology movement, gives prominence to these two points taken together, however, something quite striking emerges. The self whose destiny it is to achieve self-realisation is not the individual human being. It is rather an entity that is not properly recognised in conventional thought, the entity that consists in the human being and other elements of living nature *together*.

This appears to be at first sight an extraordinarily radical claim, one that would entail the overthrow not simply of so many of our beliefs, but our whole conception of what a belief *is*. (We would have to attribute contradictory beliefs to the same entity, for example, since when you and I appear to disagree over something these would be thoughts that in fact both belonged to the *one* thing that was nature.)

Observations like these must *surely* miss the point. The thesis that we are parts of a greater whole must surely be being misunderstood if it appears vulnerable to this kind of banality. Yet it is difficult to understand clearly how exactly it is to be understood in a more adult way. What exactly is the difference between our being individuals with intimate and intricate dependencies on nature, and our not being individuals in our own right but parts of a greater whole?

I shall try and explain later (Chapter 6) that this is such a difficult question for us in the Modern world because our 'atomistic' science doesn't support a distinction between a collection of things and a thing that consists of a collection of things – although that distinction is there in previous frameworks of thought. The fact that Modern science makes it difficult to formulate the thesis of 'unity', which 'Deep Ecology' is trying to get across, will of course disconcert it not one bit.

If its abstractions are not altogether transparent, the practical message of Deep Ecology is powerful. It has been formulated as a kind of 'manifesto' by Naess, its architect.

Box 3.3

A platform for the Deep Ecology movement

1   The flourishing of human and non-human life on Earth has intrinsic value. The value of non-human life forms is independent of the usefulness these may have for narrow human purposes.

2   Richness and diversity of life forms are values in themselves, and contribute to the flourishing of human and non-human life on Earth.

3   Humans have no right to reduce this richness and diversity, except to satisfy vital needs.

4   Present human interference with the non-human world is excessive, and the situation is rapidly worsening.

5   The flourishing of human life and cultures is compatible with a substantial decrease of the human population. The flourishing of non-human life requires such a decrease.

6   Significant change of life conditions for the better requires change in policies. These affect basic economic, technological and ideological structures.

7   The ideological change is mainly that of appreciating *life quality* (dwelling in situations of intrinsic value) rather than adhering to a high standard of living. There will be a profound awareness of the difference between big and great.

8   Those who subscribe to the foregoing points have an obligation directly or indirectly to participate in the attempt to implement the necessary changes.

(Arne Naess, *Ecology, Community and Lifestyle*, Cambridge, 1989, CUP, p. 29)

Summary

- The Romantic movement at the end of the eighteenth century developed a critique of the outlook that had become established following the scientific revolution of the seventeenth century.

- Features picked out for criticism included its commitment to the inertness of nature and the human mind, both of which it saw locked in a rigorous determinism, its lack of respect for feeling, its sustained attempt to 'explain away' some of the most distinctive features of human experience, like our appreciation of colour, taste and smell.

- We should think of the human being as capable of action and of finding expression through growth and action, and, above all, perhaps we should recognise that the distinction science presupposes between ourselves and an objective nature is a mistake.

- Deep Ecology has reinvigorated these claims, the last one in particular, and drawn up a manifesto for action.

- There are further echoes of Romanticism in the arguments I explain next. They have in common the idea that the exploitation of the environment can only be understood alongside the exploitation of other kinds – especially exploitation of women.

Questions

1   What do you take from Romantic thought that is valuable for today's problems?
2   Do you think human beings are part of nature? If so, in what sense are they free to change their ways?

Further reading

Devall, William and Sessions, George, *Deep Ecology*, Salt Lake City, UT, 1985, Gibbs M. Smith.

Fox, Warwick, *Towards a Transpersonal Ecology*, Boston, 1990, Shambhala Publications.

Naess, Arne, *Ecology, Community and Lifestyle*, Cambridge, 1989, CUP.

Taylor, Charles, *Sources of the Self*, Cambridge, 1989, CUP, Chapter 21.

Zaehner, R.C., *Mysticism Sacred and Profane*, 1957, paperback edn, Oxford, 1961, OUP.

④ The exploitation of nature and women

In this chapter I want to explore an idea that has stimulated some of the most creative thinking about our current environmental predicament. It is the thought that our exploitation of the environment is to be seen as linked with other kinds of exploitation, which are features of Western culture (at any rate), and especially with the exploitation of women.

Establishing that there are important links between the exploitation of the natural world and the exploitation, for example, of women does nothing in itself of course to exculpate science, the target of critical thought in the last two chapters, and for most writers from this point of view science remains firmly in the dock. For example, science is a kind of rape of the natural world, it has been argued, just as indefensible as any other. Science has left no life-style on the planet untouched, and even children living in the Brazilian rainforest (see Plate 4.1) alas represent no exception.

Plate 4.1 *Urueu-Wau-Wau youngsters*

Source: Colour photograph appearing in *National Geographic* 174(6), December 1988, p. 810. © *National Geographic*

From a rich field I will present just three lines of thought.

First, there is a thesis that there are distinctive virtues and strengths characteristic of femininity. These have been systematically under-valued in our culture in the past – regarded as secondary to virtues and strengths associated with masculinity – and this has in part led to the irresponsible approach, which underlies the present crisis, to our dealings with nature. If the situation is to be retrieved, it is these feminine virtues that must come to the fore.

Taking that thought one step further: It is argued that one particular weakness of the masculine virtues is the way in which they commit us to forms of social organisation that are *hierarchical*, and it is actually these that have issued in environmental exploitation. If we were guided instead by the feminine virtues, hierarchy would disappear. Instead of the mass society with which we are familiar, there would emerge small-scale communities enjoying a relationship with the environment that would be genuinely sustainable.

Second, there is the approach from a feminist starting point which aims at appropriating the powerful Marxist analysis of society – the Marxist account of what determines the features of a particular society, and what brings about historical change. According to this line of thought, Marx was essentially right in his claim that the feature which determines everything else about a society is the way it organises its production of human necessities. But what is missing is the recognition that one of the human 'necessities' is the having and bringing up of children. Introduce this recognition and you have an enriched Marxism that supports fruitful thinking about our current predicament.

Third, there is the powerful analysis offered by Val Plumwood.[1] Her thesis is that exploitation of both women and nature flow from a certain fundamental, deeply entrenched schema of thought, and that it is this that needs to be displaced.

I shall explain the latter first.

The dualisms of Western thought

There is, it is said, running through Western thought from its begin-nings to the present day, a single structural feature that is responsible for our exploitative attitude towards nature, the subsidiary role attributed to women, and for much else. This fundamental structural feature is conveniently called 'dualism'. It is a whole mode of thought. One aspect is the sharp dichotomy we make between *human beings* and *nature*, but it extends further than that. Under the grip of 'dualism' we think whenever we can in terms of two contrasting and mutually exclusive categories. We think of human beings *versus* nature, but also of reason *versus* experience, mind *versus* body, reason *versus* emotion, the self *versus* the other, animate *versus* inanimate, learnt *versus* innate and assertion *versus* negation.

Although there are very many dualistic contrasts to be found in Western thought, one of them is fundamental, and the others are variations. The foundational contrast between *reason* and *nature* was articulated by the Greeks. Since then, different dualisms have developed as different historical formations of ideas have taken shape, and, in different periods, philosophers have focused on different dualisms. Descartes, for example, attended to the dualism of mind and body. However, others have drawn attention, for example, to the dualisms between public and private, between male and female and between universal and particular.

Plumwood attributes the original dualism to Plato.[2] In Plato, she argues, there is a radical opposition between, on the one hand, the world that we encounter in our everyday existence and, on the other, a realm beyond what we can sense – a realm of 'ideas'. That is where dualism starts – with a distinction and an opposition set up between two worlds.

However, another contrast is intimately bound up with this one – a distinction between the *two human faculties* which give us *access* to the two realms.

According to Plato's account, it is our *senses* – sight, hearing, touch, etc. – that give us access to 'the world of sense'. How do we then gain access to the world of ideas? Plato's answer is: through our reason, or intellect. Our access to this second world, the world of ideas, is, Plato thinks, much more limited than our access to the world of the senses, but it is our 'reason' that reveals to us what little we can know of it.

There is thus a second 'dualism' in Platonic thought. The first is the distinction between the world of sense and the world of ideas, and the second is the distinction between sense-perception and reason.

Nature as the realm of the contingent

We must add to the definitions just given the thought that the world of sense is in some way 'inferior' to the world of ideas. Plato is often charged with holding that the world of sense is less 'real' than the world of ideas – although it is not easy to work out what this might mean. One analogy Plato himself uses is the contrast between a thing and its shadow. It perhaps makes some sense to think of a shadow as 'less real' than the object that casts it.

In Plato's thought, the status (the 'superiority' as it is often understood) of the world of ideas is bound up with his interest in the question of change. Things that were subject to change were looked upon as worthless in relation to what was unchanging.

There are two sorts of changelessness. The first is the sort of changelessness possessed by those things, if there are any, that are not subject to *destruction* or *decay* – things which will go on existing forever. And the second is the sort of changelessness that refers to a thing not *altering* – not changing any of its features. Plato's picture, as most commentators see it, was that there is a world, apart from the world of sense, inhabited exclusively by things that are not subject to change. This is his 'world of ideas'. He thought of the changeless entities that belonged to the world of ideas as immutable in both senses – they don't change their characteristics, and they are not susceptible to destruction or decay.

Plato's world of ideas is not simply a parallel world to the world of sense; the two are connected. Somehow, the items which belong to the world of sense owe their existence and the qualities they possess to the denizens of the world of ideas. When a Platonist claims that the world of ideas is 'superior' to the world of sense, what might be meant is that: The world of ideas is superior because the world of sense is entirely dependent on it.

Plato presents the picture of us living our lives out within the confines of a cave, and with our backs to its mouth. There is a fire in the cave, which throws shadows on the walls. The shadows, he suggests, are items belonging to the world of sense; the world of ideas is outside. It is clear that Plato thinks we should aspire to maximise our contact with the world outside, the world of ideas – the world of changelessness. This is a reflection of the superiority of the world of ideas.

The relative status of the two worlds carries across in Plato's thought to their respective modes of access. Just as the world of ideas is superior to the world of sense, so reason is superior to sense experience.

So we have the distinction in Plato between the world of sense and the world of ideas. And we have the claim that the world of sense is inferior to the world of ideas. Correspondingly, we have a distinction between the different modes of access to these two worlds: sense experience and reason – and the assumption that reason, by which we access the world of ideas, is superior to sense experience, which only tells of the inferior world of sense.

The consequences of this early philosophical reflection on human experience, its nature and what reality must be like in order to make it possible have – it is argued – been immense. There at the launch of Western thought, 'nature' – the world about us – has a decisively inferior evaluation put upon it. It is contrasted with something else, and that other thing is regarded as superior. Moreover our means of accessing ordinary nature, which is our sense experience, is given an inferior status, with reason taking the laurels. It is said that this perspective on nature, and on reason, once established, has conditioned Western attitudes ever since.

There are then two contrasts in Platonic thought: between the two worlds of nature and of ideas, on the one hand; and between the different means of access to these worlds, the senses and reason, on the other. What comes to have decisive influence down the centuries, according to Plumwood, is a sort of hybrid of these, a dualism between *nature* and *reason*.

I have said that the world of ideas is regarded in Platonic thought as the 'higher', more admirable, form of being (as it were) and this esteem rubs off on the means of access to it, human reason. In fact reason in Western thought is given a kind of 'hero' status, says Plumwood. The story we tell ourselves is that over the centuries our civilisation has fought for mastery over nature, with increasing and finally decisive success, and that it is reason that has been our weapon. The key concept of 'progress' in the West, Plumwood says, is 'the continual and cumulative overcoming of the domain of nature by reason'.[3]

Cartesian dualism

Though Plumwood is keen to stress that the dualism between nature and reason is rooted in ancient thought, she thinks it took on a fresh lease of life through the emergence of modern science. It assumed the form at that stage of a dualism between mind and physicality: the category of reason is 'morphed' into the category of the mind, and 'nature' stands now for the material world.

This particular incarnation, the dualism between mind and the material world, had a specially significant influence for our Modern dealings with nature. It puts all the qualities that flow from the possession of a mind on one side of the divide. Nature, which has none of these, becomes conceived of as being inert, purposeless, lacking in any kind of feeling or sensitivity, 'bereft of qualities appropriated to the human'.[4] In short, nature becomes the mechanistic universe of the Enlightenment.

Box 4.1

Nature as mechanical

Mechanism then involves a stripping process, the stripping out especially of mindlike qualities such as agency and goal-directedness. The same stripping process is applied in the case of the natural world to yield the account of nature as a machine. Nature is taken to have no originative power within itself, and to be devoid of teleology, to be 'plastic'. Cosmos and organism emerge as a meaningless assemblage of parts because their organising principles are lost in the destruction of intentional description, and its isolation in the separate organ of mind.

Consciousness now divides the universe completely in a total cleavage between the thinking being and mindless nature, and between the thinking substance and 'its' body, which becomes the division between consciousness and clockwork. Gone is the teleological and organic in biological explanation. Mind is defining of and confined to human knowers, and nature is merely alien.

(Val Plumwood, *Feminism and the Mastery of Nature*,
London, 1993, Routledge, pp. 115–16)

Thus, at the level of the person, the Cartesian dualism has a mind as a quite different type of thing from the body, but superior to it, and 'driving' it, so to speak. While, at the level of the world as a whole, there is a sharp distinction between mind and mechanised nature, with the inferior status of the latter making it vulnerable to any exploitation that the superior mind chooses.

Plumwood identifies many other dualisms – see Box 4.2 – and seeks to represent them all as transforms of the same fundamental. Of the elements listed on the left-hand side, she says, virtually all of them can be represented as forms of reason, while virtually all on the right-hand side can be represented as forms of nature. Together the sequence of dualisms illustrated in the list form something like a geological fault-line in the structure of Western thought.

Box 4.2

Key elements of the dualistic structure in Western thought

These key elements are the following sets of contrasting pairs:

culture	nature
reason	nature
male	female
mind	body (nature)
master	slave
reason	matter (physicality)
rationality	animality (nature)
reason	emotion (nature)
mind/spirit	nature
freedom	necessity (nature)
universal	particular
human	nature (non-human)
civilised	primitive (nature)
production	reproduction (nature)
public	private
subject	object
self	other

(From Val Plumwood, *Feminism and the Mastery of Nature*, London, 1993, Routledge, p. 43)

All these descendants of the Platonic archetype, says Plumwood, inherit the original valuation of reason above nature. Civilised on the left-hand side is better than primitive on the right, public better than private, rationality than animality, freedom than necessity, and so on. You have not just a fault-line, but a division that separates better from worse, superior from inferior.

It is this interlocking structure of dualisms that defines for the contemporary world what it is to belong to nature. It is to be defined as:

> passive, as non-agent and non-subject, as the 'environment' or invisible background conditions against which the 'foreground' achievements of reason or culture...take place. It is to be defined as a...resource empty of its own purposes or meanings, and hence available to be annexed for the purposes of those supposedly identified with reason or intellect, and to be conceived and moulded in relation to these purposes.[5]

So far Plumwood offers an account of the nature of our contemporary attitudes to nature, which she thinks derives from an archetypal dualism established by Plato. But she also thinks (as is apparent from the list) that the dualism between nature and reason has produced as one of its offspring a dualism between male and female, and it is in this way that, for Plumwood, the exploitation of nature is linked with the exploitation of women. By and large, the qualities listed on the left-hand side are ones that are 'traditionally appropriated' to men, while the right-hand side are qualities 'traditionally appropriated to women'.[6]

Feminism and Marxism

The perspective on human social life known as Marxism is still the beacon that draws those who want to articulate their radical dissatisfaction with the way things are. They don't always agree with it, of course, but they very often define their position by saying how they differ from it.

Marx (1818–83) argued that the fundamental feature of a society was the means it adopted to produce the necessities of life – how it organ-

Box 4.3

The gendered dualism of reason and nature

A gendered reason/nature contrast appears as the overarching, most general, basic and connecting form of these dualisms, capable of nuances and inflections and a great variety of elaboration and development.

(Val Plumwood, *Feminism and the Mastery of Nature*, London, 1993, Routledge, p. 44)

ises itself to produce food, clothing, shelter and so on. A society's 'mode of production', he thought, is fundamental in the sense that from this one thing flows everything else. If you have your necessities produced by slave labour, you will have a particular form of social life and a particular form of culture, including whatever you have in the way of art, writing and philosophy. Alternatively, if your mode of production is based on individuals relying on markets, different forms of social structure and culture will result. Marx's analysis implies that if we need to alter our manner of relating to nature, this will only come through a change in our fundamental economic structure.

That said, Marxism has to dig deep in its intellectual resources to offer much of an answer to the problem identified by environmentalists, because – in conception at any rate – Marxism doesn't really recognise what the modern world does to the environment as exploitation. Industrial production, against whose door contemporary environmentalism tends to lay a good deal of responsibility, was for Marx the key to the future. It represented for him the potential to release human beings from drudgery, and so make possible the integrated, creative life he thought human nature needs for its fulfilment.

Critics who are basically sympathetic have sometimes asked if Marx was right to identify human 'necessities' as he does. What exactly counts as a 'necessity'? – This is a question that needs clarifying before we can know exactly what is involved in the claim that producing 'necessities' lies at the foundation of human societies. They have argued that, when Marx refers to the growing of food, and the making of clothing and shelter as foundational, he is 'back-grounding' something that appears overwhelmingly obvious and significant when attention is drawn to it, namely the need to 'produce' life itself. The activity of re-production is surely the most fundamental human activity of all. As in fact Marx's collaborator Friedrich Engels points out: if we speak of production we must include not only 'the production of the means of subsistence' but also 'the production of human beings themselves'.[7]

This allows a feminist appropriation of Marxism, which creates a distinctive perspective on our environmental problems. To solve them, it argues, we need to alter our 'means of producing our necessities', but this must be understood as involving a revolutionary change in our approach to reproduction.

Cultural ecofeminism

Cultural ecofeminism recognises a fundamental difference between men and women: that women are closer than men to the nature to which they both belong. Women are more attuned to the needs of fellow human beings and better equipped to care for them; they are more responsive to others and to nature, both emotionally and spiritually. So far, the cultural feminist accepts the view of femininity that enjoyed a reputation for being enlightened and progressive in times before feminism took its modern form – the view, very broadly, that was to be found articulated by the more perceptive cultural critics.

What is accepted then by cultural feminists is that women and men have different characters, but they add to this that those associated with woman have in times past been systematically undervalued. This is what has been wrong: not the identification of distinctively feminine characteristics but the perspective on these that treats them as secondary. The emotional responsiveness, the urge to care and nurture, spiritual awareness – these are characteristics that are both distinctive of woman, insist the cultural feminists, and at the same time absolutely essential to a defensible and sustainable way of life.

Ways of life might be imagined in which distinctively feminine virtues are given their proper place and respect, and several writers look to history to find actual forms of life that approach this ideal. They have to look back a good distance, however. As Carolyn Merchant reports: 'Many cultural feminists celebrate an era of prehistory when nature was symbolised by pregnant female figures, trees, butterflies, and snakes and in which women were held in high esteem as bringers forth of life.'[8]

Critics say there is a sense in which, on the one hand, this absolves men of responsibility: it appears to be not their biological business to care for the world. On the other hand (because it ignores Marxism), it stands accused of having no very hard-edged account of how women might achieve the power needed to assert their priorities.[9]

Social ecology

While Marxism looks to the huge productivity of large-scale industrial processes to free human beings from the need to labour, others have taken the view that you can't have large-scale industry without elaborate social structures to support it. For example, a large factory needs raw materials supplied according to a regular schedule, and this requires a high degree of complexly interacting, reliable behaviour on the part of a large number of people. These imperatives of large-scale factory production are not consistent, it is argued, with the enjoyment by human beings of what is fundamentally important to them – their freedom. To live in a society based on large-scale industrial production is to be subject to a tight web of constraints and pressures.

So although Marx himself attached the highest importance to freedom, and thought this would be realised with the potential of large-scale industrial technology, others have seen an incompatibility there. But suppose, they have suggested, we develop a high-productivity technology that could operate on a *small* scale? Would that not lift the constraints? A small-scale but highly automated system of production could be used to support communities that were essentially small in scale themselves. All of the limitations on freedom and all of the subjugation of the individual, which come with the 'mass' society, would be eliminated, transcending even the struggle to survive that has (allegedly) characterised small communities in the pre-industrial world.

A return from the mass society to much smaller 'human-scale' communities has long been advocated by *anarchists* – those thinkers who, one might say, put human freedom beyond compromise. Not all have looked to the potential of high technology to make life in such settings materially comfortable. Others have thought it better to develop what is known as 'intermediate' technology. These would be techniques that not only lack 'high-tech' productivity, but also the high-tech impact on the environment, and the high-tech demands on the human beings who depend on it.[10]

This approach to the question of how we should organise our life together is sometimes known as *social ecology*. Exploring it takes us into political philosophy, which we are treating (for reasons solely of space, not relevance or importance) as outside the scope of this book – except for some sketching in Chapter 6.

One attraction of social ecology to (some) feminists is its promise of a less exploitative relation to nature. In the mass society we belong to, nature can be over-exploited without that fact being immediately obvious. A pleasant local situation may be being bought at the expense of unrecoverable devastation elsewhere that is out of sight. In the end the chickens will come home to roost, but people at large, if their circumstances are comfortable, may allow themselves to ignore or deny this inevitability. If, instead, the human community one belongs to is *small*, as well as self-contained, the consequences of destructive practices will be apparent to everyone. It is argued that the necessity for finding ways of growing and making things that are sustainable into the indefinite future will be apparent to everybody.

But there is another, more fundamental, thought connecting social ecology with feminism. It is the claim that environment-damaging practices are in fact *grounded* in hierarchical social structures.

Anarchists favour small-scale communities just because they see them as free of the power hierarchies that otherwise destroy human freedom. If it is power hierarchies that ground damaging environmental practices, the kind of life envisaged by social ecology can only be more sustainable.

Summary

- There are plenty of echoes here of Romantic thought. It does not require a tremendous leap of the imagination to see the exploitation of Mother Nature in our culture as part of the exploitation of women in general. Some have suggested that if the character of a woman is more 'nurturing' then that is exactly what our culture needs to adopt in its dealings with the environment. Others have turned to that classical analysis of exploitation, Marxism, for a basic perspective on society and its workings; Marxism, which some have amended so that a society's arrangements for the creation and education of its next generation are counted as part of the 'means of production'. Social ecology places its faith not on the productivity of the modern industrial plant but in the break-up of mass society into small self-sustaining groups.

- Plumwood's argument is that exploitation of both women and nature flow from a certain pattern of thought. It is a pattern, dualism, that was laid down right at the birth of Western civilisation, and, though taking a variety of forms across the centuries, with its fundamental oppositional categories of reason and nature, it remains essentially in place.

- The target of much of the foregoing, in this chapter and the others, has been, in one way or another, *science*. I want now to present what can sometimes be seen as an alternative to science – or if not to science as such, to science as it has become established – or if not to the whole of conventional science at least to important regions of it. *Phenomenology* has its roots in a different tradition from the one within which my discussion so far has been cast, and it is one that I assume will be less familiar. To do it justice, I have asked a colleague to present it, and offered the space to allow something of the flavour as well as its interest from the environmental point of view to be conveyed.

Questions

1 What is the significance of the phrase 'Mother Nature'?
2 To what extent are environmental problems to be put down to the scale of human society?

Further reading

Colard, Andree and Contrucci, Joyce, *Rape of the Wild*, Bloomington, 1989, Indiana University Press.

Griffin, Susan, *Women and Nature*, New York, 1980, Harper & Row.

Keller, Evelyn Fox, *A Feeling for the Organism*, San Francisco, 1983, W.H. Freeman & Co.

Marx, Karl, *Selected Writings*, ed. David McLellan, Oxford, 1977, Oxford University Press.

Merchant, Carolyn, *The Death of Nature*, San Francisco, 1989, Harper & Row.

Plumwood, Val, *Feminism and the Mastery of Nature*, London, 1993, Routledge.

Warren, Karen J., *Ecological Feminist Philosophies*, Bloomington, 1996, Indiana University Press.

5 Phenomenology and the environment

The phenomenological philosophers believe that philosophy, at least since Descartes, has misunderstood the relationship – what it is, and what it ought to be – between humans and the rest of the world. Their aim is to explain how the misunderstanding arose and to restore a proper understanding of the world and our place in it.

The crisis

Edmund Husserl (1859–1938) thought that there was a crisis in European thinking. The crisis he meant was not our current environmental crisis, but had he survived to see our present predicament he would doubtless have thought that the two crises were closely linked, the one he wrote of being the root of the present one.

Husserl's claim was that, with the rise of modern science from Galileo and Newton, scientific concepts and theories had become highly abstract, divorced from the actual experience from which they are abstractions, and in need of 'anchoring' in experience. The task he set himself – and it is a philosophical not a scientific one – was to explore experience, the 'lived world' of which science purports to give a general picture, and discover the laws explaining how it works.

This is not a new philosophical enterprise, as Husserl was well aware. In the seventeenth century, as explained in Chapter 2, thinkers such as Locke and Descartes had taken the problem to be this: the workings of the physical world, and that includes the workings of human bodies, can all be explained in terms of size, shape and motion. But the world as we experience it has other properties: it is noisy, colourful and it has smells and tastes.

A dominant metaphor was that these properties are not properties of objects but are 'contents' of the mind. This clearly *is* a metaphor since, literally speaking, the mind has no extension and so it cannot have any contents. This metaphor of the mind as a container was used in psychology. The contents were said to be 'sensations'.

Husserl's target was not physics as such: the abstractions there are useful abstractions. Psychology, in contrast, he believed had gone badly wrong in its adoption of this metaphor and its belief in sensations as the basic units, the building blocks, of human experience. A later phenomenologist, Merleau-Ponty (1908–61), offers a detailed criticism of scientific psychology based on the notion of sensation. One important criticism is that psychology so based has failed on its own terms to explain the workings of the mind. The notion of a sensation as the basic unit of experience raises problems within psychology that psychology has not answered satisfactorily. To this charge, the psychologists could legitimately reply that there is much explanatory work yet to be done, but this is no grounds for giving up the task as they conceive it of explaining how perception, knowledge and action are 'constructed' out of sensations, the basic data or input of experience. But this response fails to meet a deeper challenge that Husserl (and Merleau-Ponty) make to psychology. This is, not just that psychology so far has had rather limited explanatory success, but that it cannot possibly succeed in explaining human knowledge and action because it is based on an erroneous account of what the basic data, the input into the human mind, is.

The phenomenologists' diagnosis of this mistake in interpreting the basic data of consciousness is that psychology, rather than explore what it is like to be conscious, has drawn conclusions – based on what physics tells us the world is like, and what biology tells us the sense organs are like – about what basic conscious experience *must* be like. The task of phenomenology is to show that this conclusion is false.

Phenomenology then presents science in a radically unorthodox light. Science, with its mechanistic account of how the world works, has had great success. But, in what does this success consist? The orthodox view would say the success consists in the truth of science. Phenomenology, in contrast, especially Heidegger (1889–1976), would say that the success lies in the fact that science is an instrument that enables us to manipulate and control the world technologically.

A familiar view of the relation between science and technology is that there is pure science, which technology applies the findings of. Heidegger rejects this orthodox view. He argues that science is through and through technological; that its abstraction, laws, concepts, and its dominant metaphor of nature as a mechanism are all guided by and geared to the aim of technological control over nature. This being so, science is not value free, instead being laden with this value of humans as controllers of nature. To regard nature as mechanistic is to regard it from the very start as like an artefact, a machine, or something we have made. But we have not made nature; rather, it has made us. This is perhaps the true message we need to get from evolutionary theory. So, for Heidegger, science is not a purely factual matter. Scientific theory and practice take for granted certain values. Are they values which we want to hold? In the light of the environmental crisis, we might well reply negatively to that question.

The modern world-view is that science tells us what the facts are. What use we make of the scientific information depends on what our values are. Values are what we determine and then project onto the world. Phenomenology argues that science already embodies a system of values, of the world as a commodity to be used by us however we want to, as having value only in so far as we can use it as an instrument to our ends. Phenomenology further argues that this world-view represents a 'falling' from a richer way of being on our part. Via this world-view, we have lost sight of something deep and precious in the way we are essentially situated, at home in the world.

The task is to weed out and challenge the screen of utilitarian values and concepts that distort our experience of the world. We need to take a fresh look at the world and at ourselves. We are not, as Descartes had it, pure consciousnesses that happen to be lodged in mechanical bodies. We are essentially embodied, essentially situated in a world of significance and value. Modern life may cover this up, but it still underlies modern 'alienated' living, and close scrutiny and sensitive description can reveal the true nature of our 'being-in-the-world', our life-worlds that are not devoid of values.

The life-world

The life-world, the world as we live in it rather than theorise about, is one in which subjects and objects are essentially related to each other. Each activity in the life-world of a subject involves interaction with an object, and all objects in the life-world are objects *for* the subject. That all consciousness is intentional is a leading thesis of phenomenology. But this thesis of intentionality is properly construed only if it is seen as a feature of the life-world. Intentionality is that feature of consciousness whereby consciousness is conscious of something. Brentano (1838–1917), a founder-figure of phenomenology, held that this was the defining property of the mental: all mental states are directed towards, about, of, objects. He held that no physical state had intentionality: the colour of a tree or size of a mountain is not about or directed towards anything (see Box 5.1).

Box 5.1

Intentionality

If you think about one or two examples of sentences, you will notice that they seem to be *about* things.

'George Fox was a religious man' seems to be about George Fox.

'Charles Dickens wrote *Dombey and Son*' seems to be about Charles Dickens writing the novel.

'I have this troublesome toothache' seems to be about me having a toothache.

The claim that sentences are *about* things is expressed by saying that they have 'intentionality'.

Mental states appear to have intentionality too. Take, as an example, one of the thoughts that may go through my mind as I watch the *Batman* film: the thought that my brother lives in London.

Here again is something that seems to have 'aboutness'. It seems to be about my brother and his living in London.

There is on the one hand my thought. And on the other a fact.

And the thought seems to be 'about' the fact.

I have just said that intentionality is 'aboutness'.

But being 'about' something is perhaps too broad a notion when it comes down to detail.

You might say that a tidal wave was 'about' the earthquake that produced it. Here you have just a causal relationship between two things. This *isn't* the 'aboutness' that is intentionality.

Just think of the sense in which a thought is *about* something. (My thought that my brother lives in London.) When I say this thought is about my brother's being in London, I don't just mean that my brother's being in London brought it about. I mean my thought somehow in itself *points* to a certain state of affairs, namely my brother's being in London. It points to this state of affairs in a way that the tidal wave does not point to the earthquake that generated it.

Brentano introduced the idea that intentionality is what makes thought what it is. Everything in the realm of thought was intentional, he maintained, and it was simply this intentionality that made it 'mental'.

Even pains are thought on this theory to have intentionality. Suppose I have toothache. Does this ache 'point' to something beyond itself in the way that my belief that my brother lives in London points to something beyond itself, namely a state of affairs: my brother's living in London? Yes, this theory maintains, it does. The ache that I have 'points' to my tooth.

Phenomenology maintains that all mental states are *directed* – they *point*. This is sometimes put by saying that all mental states are directed towards 'objects', though the sense of 'object' here is perhaps unusual.

So, for phenomenology, conscious subjects depend on 'objects', and objects are first and foremost objects for conscious subjects. The objects of which science speaks, since science is based on human observation, are, or at least start out as, objects of consciousness.

It is part of the crisis that phenomenologists identify as afflicting our thinking that intentionality has not been properly understood. The intentionality of consciousness has been a problem since Descartes. Having divided the world into external objects and pure consciousness (see Chapter 4 above), the question arises: how can the latter be 'about' the former? The only available answers are that the objects cause intentional states or that the intentional states represent objects. But in neither case can there be any conceivable guarantee that the objects exist or that the representations are accurate, because all consciousness can ever encounter are the effects of the alleged causes or the representations of the alleged objects.

Another problem related to the problem of intentionality is the problem of value. Since, post-Descartes, the world is regarded as a purely factual realm, values cannot come from objects in the world. So they must come from subjects – we are the source of values, and we project values onto the world.

But this, claims phenomenology, is a theoretical stance upon the world and ourselves. If we explore the lived world, we find that we cannot separate out so easily subjects from objects, values from facts. The Cartesian world-view has prised them apart and worried ever since how they could be united. Why, the phenomenologists ask, prise them apart in the first place? They are related through and through. Moreover, if the relations that hold in the life-world are properly investigated, it will emerge that neither subjects or objects, nor facts or values, are as presented by the Cartesian model. Phenomenology takes intentionality as fundamental: subjects and objects are first and foremost related to each other. Any account of either must pay heed to this relatedness.

The life-world is not one in which we gaze at value-free objects and decide on values, in the light of preferences or judgements formed by self-contemplation in isolation from those objects. In the lived world, objects are experienced as inviting, repellent, frightening or horrible, and preferences and judgements are formed on the basis of how the objects have affected us and what we have learnt to do with them.

It should already be beginning to emerge that phenomenology promises both a critique of much environmental thinking and an alternative, more eco-friendly, world-view.

The reduction

Phenomenology aims to reveal essential features of this life-world or lived world. The key notion is that we need to adopt a reflective standpoint from which we can take a non-theoretical, unprejudiced look at our experiences. The prejudices we should seek to avoid are the ones that, according to Husserl, have given rise to the crisis. Overall, the claim is that, if we attend to our experience, our 'lived world', we shall find that our experience is not one of passively receiving sensations caused by a mechanistic world, but one of interacting with a valuable and meaningful environment. The phenomenological aim is to describe these interactions.

Different phenomenologists recommend different ways of achieving this reflective standpoint. Husserl recommended that we put aside – 'bracket' – our everyday belief that the world exists. This, he believed, would enable us to focus on and describe the world as we experience it. Husserl believed that we, as subjects, can take a stance that 'transcends' the world of objects, hence his phenomenology is 'transcendental'. Existential phenomenologists believe that no such transcendental standpoint is achievable. Heidegger believed that the reflective standpoint was not so easily achieved. The task of phenomenology, he thought, is to allow things to show themselves. But they cannot do this as long as we hold our everyday prejudices that 'alienate' us from the world. Heidegger's work was devoted to exposing these prejudices. Merleau-Ponty believed that Husserl's bracketing, also called 'the reduction', could never be more than partial. We can never put aside all our assumptions about the existence of objects. We may focus on some aspect of our lived world, but we shall still be taking for granted other aspects.

What all three thinkers have in common is the belief that in ordinary living we ignore important features of our interactions with the world. These interactions, and the true nature of the participants in them – human subjects and the world of objects – are implicit in ordinary experience and can be made explicit from a suitable phenomenological standpoint.

If, as is claimed by some environmental philosophers, in modern life we have lost sight of what our proper relationship with our environment is, then phenomenology promises a way of bringing this into view again. In particular, the claim would be that our lived worlds incorporate all manner of activities that are constitutive of valuing it. A proper account of valuing can result from phenomenological enquiry.

Phenomenological description

A phenomenological field trip

Phenomenologists insist that one should describe particular, concrete, actual experiences, and not imagined, remembered or possible ones. Just as some nature painters insist on painting from nature, so phenomenology of nature should be done from nature. So, go out and find yourself something to describe. You will probably be surprised at what hard work it is and at what emerges.

The aim is to uncover the experiential basis of anything we might say about what we are describing. If you are a natural scientist, you will likely begin your description using classifications – there is an oak tree, a limestone pavement, a wood pigeon, etc. What enables you to classify things in these ways? How do you identify the items you classify? You may well have forgotten your original training in the field. It might be helpful to take someone with you who doesn't know an oak tree from a beech. What do you have to point out to them? What do they have to learn to distinguish to acquire field skills?

What will emerge, if you are patient, will be much more than you expect.

You recognise this sort of territory as the sort of place oak trees grow. How? What are the signs?

You spot a tree in the distance – it looks like an oak. So what shape is it? Are all oak trees *exactly* that shape? Are no other trees that shape? So how *do* you know it's an oak? Is there something distinctive about the colours? Is it moving in the breeze in a distinctive way? Would anything convince you that it wasn't an oak? For instance, walking round to the other side of it and finding it was a cleverly made sculpture. So, you have expectations of what other experiences of it would be like, how it would feel, what its other sides are like, and what its leaves look like. These expectations are what Husserl calls 'horizons' of experience. What seems like a simple sight of a tree turns out to have 'implicit' in it all manner of other possible experiences and expectations, without which you wouldn't be able to identify this as an oak tree.

You notice an acorn. Clearly a sign of an oak tree. But what is involved in 'reading' that sign? You know it is a nut, the fruit of the oak tree. Recognising that involves knowing what nuts are. It also involves knowing that trees bear fruit that fall off and produce other trees. Probably you know that squirrels eat acorns, and that acorns rattle when you shake them. All of this is involved in your recognising an acorn.

You get closer to the tree and see its leaves. They have a very distinctive colour and shape. But would you be able to draw the shape, or pick out the colour from a colour chart in the absence of the tree? Quite possibly not – your skills of identifying the leaf are activated by the whole experience.

What you are aiming to do is to reveal how much is involved in a seemingly simple experience, the recognition of an oak tree. Normally, we take such an experience for granted. The phenomenological aim is to stop taking it for granted and describe all the knowledge and perceptual and motor skills involved in even the simplest experiences of the world.

Notice, in particular, how much of your description will be about you and not just about the oak tree. Your experience involves your expectations, beliefs and abilities. Scientific classifications, which make no reference to these aspects of experience, are abstractions from the experience: they have abstracted the subject from the experience; abstracted from the lived world.

If you are an artist, you will find other things to describe. The graceful curve of the branches, the grandeur of the tree, the insignificance but enormous potential of the acorn, the play of light through the leaves, the dappling of the ground beneath. How do you recognise these things? – the grace, grandeur, play? How would you teach someone who didn't have your aesthetic sensibilities to recognise such things?

Suppose they don't see the grace. Maybe they need to focus on just one branch, not the tree as a whole. Perhaps you then move your arm in a graceful way, making something like the same sort of shape the branch makes as it moves in the wind. You encourage them to move in that way, to feel the difference between that and jerky, graceless movements. What would they find graceful? A ballet dancer perhaps? How do her graceful movements resemble the graceful branch? In this way we uncover the basis of our aesthetic beliefs about nature. What we find is implicit in our experience: what we have to learn in order to have the experience. An ornithologist, a surveyor and a historian might have different stories to tell, different skills to impart, different riches to uncover. Experiences are revealed, in this way, to be much richer and more complex than one at first supposes.

Let us now reflect on this field trip. What have we done? First we selected what Husserl calls a 'theme': encountering an oak tree. Second we undertook to describe the phenomenon of encountering an oak tree. Phenomena are events in the life-world, all involving both subject and object. Third we described the actual experiences and activities involved, and the features of the tree and other objects noticed in the overall phenomenon. Fourth we described the 'horizons' of the phenomenon. These are things like expectations, background knowledge, motor skills, and capacities to recognise. These are implicit rather than explicit in our encountering the oak tree. They are hard to tease out because they are so familiar, so much taken for granted in our ordinary, everyday life, that we don't notice them.

There are two problems here. One is that focusing attention on these taken for granted skills and capacities can be disabling. We may feel like the centipede who was walking just fine until asked which leg he put forward first. So, in our case, one activity involved in our encounter is focusing the eyes on the tree at various distances, then on various aspects or parts of the tree. But, paying attention to how one focuses one's eyes tends to make it difficult to do, it begins to feel unnatural. The second problem is that attending to the very familiar events in one's life, revealing their complexities, can lead one to present them as extraordinary or surprising. But this would be to mis-describe them because they are, as events in one's life-world, not surprising or remarkable.

Husserl described phenomenology as a 'rigorous discipline'. One thing he meant by this is that phenomenological description is very hard work. These are two things that make it hard work. It is in some measure comparable to the hard work involved in drawing. One's attention has to shift constantly between the object and the drawing.

So with phenomenological description, one's attention has to shift constantly between the phenomenon and the description. And, as with drawing, it brings to light all manner of things one did not originally notice.

There is another difficulty in producing phenomenological description. This takes the form of not knowing how to start or where to stop. Descriptions usually have some purpose – one is giving directions, explaining why one left, etc. What is the phenomenological purpose? The effect of phenomenological description can be to enhance one's sensitivity to and appreciation of one's environment and how one is attuned to it. But if that effect is adopted as a purpose, a goal, the description might go awry.

Philosophically, phenomenological description has two purposes. One is to undermine a philosophical orthodoxy that is taken for granted in much of what we say. The second is to uncover essential features of the lived world. Let us look at each of these in turn.

Phenomenological description should be without presupposition or prejudice. One presupposition or prejudice is that there is a sharp distinction between objective descriptions of the world and subjective descriptions of our responses to it, experiences of it or beliefs about it.

Over many years of teaching phenomenology and trying to encourage students to produce phenomenological descriptions, I have observed people experiencing the difficulties of dropping these prejudices. As long as people are describing objects, they are confident, especially if they have a scientific training with respect to what they are describing. They feel they know what they are talking about. Others less well informed listen to them, and defer to their superior scientific knowledge. Unfortunately, these are the people who can deviate rather quickly from a phenomenological description. They take scientific classifications for granted, rather than exploring the experiential basis for those classifications. They refer to objects, and forces that are not actually perceivable. They appeal to general laws that explain the phenomenon instead of seeking simply to describe it.

In contrast, when the phenomenological description is concerned to describe the involvement of the subject in the description, either the psychologists take over and, as with the physical scientists, offer classifications and explanations rather than descriptions – this often takes the form of explaining the phenomena away. Alternatively, people feel they are talking about themselves and this induces either acute embarrassment or verbal excess. Both of these spring from the same belief: that here anything goes, one can say whatever one likes, there are no rules.

So, are there rules and if so what are they? There are two rules of description. First, every description of an object experienced must be accompanied by a description of how the subject is experiencing it and every description of an experiencing subject must be accompanied by a description of the object being experienced. Second, one must hold firm to the phenomenon one set out to describe: one's description must stick to what is integral to experiencing the object in question. Being over-theoretical takes one away from one's experience; being over-fanciful takes one away from the experienced object.

The best advice at this stage is probably to say don't be embarrassed. Choose a phenomenon in which you are sufficiently interested to want to think hard about, to return to it again and again to articulate exactly what it is about the phenomenon that gives it the 'draw' it has for you: experiencing a spectacular sunset, a disused barn, a stretch of canal, a wild animal, and then describe away. It is really at the next stage of phenomenology that the description is scrutinised and some bits of it might have to go.

The life-world is not merely factual, so do not avoid evaluative terms in your description if they seem appropriate – use metaphor, aesthetic terms or emotional language if it seems right. Don't be deterred by the thought that what you are saying is 'merely subjective'. That challenge is a remnant of the presupposition you are supposed to be dropping. If you think that what you are describing is an integral part of the phenomenon, then put it in. Phenomenologists believe that phenomenological description shows the presuppositions to be false of the life-world, so you must not be deterred from including something in your description on the grounds that it goes against the presuppositions.

The trouble with just saying 'describe away' is that one either can't get started at all, or, having started, one cannot stop. It may be that knowing where the description is heading will help in getting started, but it can be a danger – the description can be very 'thin' and so miss its mark if the description is too goal-orientated. The goal, the second purpose of phenomenological description, is to reveal essences. How is that done?

Essences

The search for essences is the next, more theoretical, stage of phenomenology. The aim is to find essential features of phenomena, features that make them the phenomena they are. In selecting a phenomenon, one had what Husserl calls a 'theme' in mind. One selected that concrete phenomenon as a phenomenon of a certain kind, an exemplar of a theme. Suppose your concrete phenomenon was swimming in a lake. What is your theme? It might be swimming in any water. It might be swimming in a lake in contrast to the sea or a pool. It might be experiencing a lake, not necessarily swimming – boating or walking round would equally be exemplars of that theme.

When you search for the essence, what is essential to your description, what you are looking for are those features of the phenomenon without which it wouldn't be an exemplar of that theme. Having got the phenomenological description – revealing, making explicit, what is implicit in the phenomena – we want to sort out what is essential and what is coincidental in the description. The move to essence seeks to avoid speculation, mind-wandering, or idiosyncratic associations that might get into the description.

Phenomenologists differ on how to discover essences. Husserl's method for finding essences is to work through a description systematically. For each element of the phenomenon you have described, you must try to imagine the phenomenon without that element. If what you imagine is still an exemplar of your theme, then the element is not essential to the phenomenon. If what you imagine is not an exemplar of your theme, then the element is essential to the phenomenon. For example, your encounter with the oak tree might have set your mind wandering off into a personal reverie about a childhood picnic under a similar tree. If you included that in your description, this is where it gets deleted. The oak tree happened to remind you of that occasion, but it might not have. You can imagine encountering the oak without indulging in the reverie. The reverie is not an essential part of the phenomenon. Your expectation about what the tree would look like from the other side, however, is something that you cannot imagine an encounter with an oak tree lacking. So that is part of the essence of the phenomenon.

Both Heidegger and Merleau-Ponty found this aspect of Husserl's method unsatisfactory. They both thought that the discovery of what is essential in the lived world is a more complex task. Heidegger believed that we need to strip away all of our ideas and concepts that he believed normally 'distance' us from the world. Once this has been accomplished, the world of things, what Heidegger calls 'Being', would 'show itself'. 'How would we know when we had succeeded?', would be the obvious question to ask here. Heidegger gives no clear answer to that question. What he does, though, is to point to all sorts of ways in which our dealings with the world do involve some sort of 'distancing', alienation or inauthenticity. These are easier to recognise than the pure encounter with Being with which Heidegger contrasts them. An example would be, on encountering our oak tree, regarding it only as potential shelving. Or, in case you're getting bored with the oak tree, think of the tourist, so jaded by the trip, that each view becomes just another lake and mountain, each work of art just another old painting. There is clearly something wrong with these responses. They involve a failure to experience what is there to be found. Much of modern life can seem to be like this. In our rush to get to work, we see only the car in front and miss the stunning sunrise; in our haste to catch a train, other people are just a queue. If we made ourselves aware of our alienation, we might better appreciate our environment. And if we appreciated it better, we might come to treat it better. That was certainly Heidegger's hope, though optimism about his fellow man was scarcely his strong point. Heidegger then gives us no clear-cut way of revealing essences, but he does offer a thought-provoking way of detecting the inessential with which we are so often concerned in our busy lives.

Merleau-Ponty sought to 'put essence back into existence'. One thing he meant by this was that Husserl's procedure for discovering essences by using imagination would not necessarily give the right answers. Pure imagination can lead us to ignore how very much of our existence is embedded in the world. For example, take the phenomenon of going to bed in your own familiar house. You might imagine that it is not essential to this operation that the light switches are located just where they are: you could, after all, have them moved and you would adjust to the new places. Merleau-Ponty would recommend not just imagining the change but doing it. One does eventually adjust, as one eventually adjusts to a new car or new computer keyboard. If you make such changes, you will notice how your hands just move to the old position of the light switches, even while you are telling yourself that they are not there. Actual changes rather than just imagined ones bring to light much better the essential elements in our practical dealings with our familiar worlds. You can reflect on moving around your house and imagine that switching the light off just there is not essential. But move the switch and you will find your hands groping, your body disorientated, your mind distracted. The movement of the switch has changed the familiar phenomenon under scrutiny: that precise bodily movement was an essential part of it.

Ontology

Phenomenological description can expose a wealth of ways in which we relate to the world. Can it say anything general about the relationship? Husserl indicated that he thought it was not value-free. Merleau-Ponty indicated that wonder in the face of the world underlies all our more practical dealings with it. It is Heidegger who perhaps is most explicit about what our fundamental, essential relation with the world of objects is. It is one of care, of mindfulness, of dwelling. Our fundamental relation with the world is that we value it; or, rather, our ways of relating to the world involve the activities that make up what valuing is. Notice the shift from 'value' as a noun to 'valuing' as a verb. If we use the noun, we expect a thing that we might find. The verb suggests activities that we might engage in, or relationships we might forge.

In the debate concerning what kind of value the environment has, the chief dispute is between those who claim that the environment has instrumental value (i.e. its value lies in its ability to serve our ends), and those who claim it has intrinsic or inherent value (i.e. it has value in its own right and so should be respected whether it serves our ends or not). Phenomenology offers a way of exploring both of these positions. What are the phenomena involved in using the environment as a means to our ends? What are the phenomena of discovering and respecting values in the environment? If we explore those phenomena, we might develop a richer account of what this activity of valuing involves, how we use or abuse our environment, how we

respect or disrespect it, and what in it we use, abuse, respect or disrespect.

So, the relation between subjects and objects is not value-free. It is
rather what makes up the complex activity of valuing. Subjects and
objects are internally related. That is to say, we do not depend on
objects just for sustenance and entertainment. We are as we are
because of the world we are in. We could not use, abuse, respect or
disrespect if there were no world. We could not have the ends we do
or use the means if we were not situated in the world. Any attempt to
describe the life of a subject without reference to objects, or objects
without reference to subjects, will leave out of the account what is
involved in valuing, just as describing members of a team without
reference to the team will leave out of the picture and make inexpli-
cable notions such as team spirit, loyalty, co-operation or 'being on a
roll'.

The subject

The emphasis phenomenology places on the intimate relation
between subjects and the world of objects is clearly relevant to environmental philosophy. The characterisations it gives of subjects and
objects, based on the fact of their fundamental relatedness, similarly
can be seen to shed new light on environmental issues.

Husserl's notion of a subject that 'transcends' the world of objects is
perhaps of little significance environmentally. It might be taken to
support those who believe that we can survive as brains in vats or as
computer programs. However, there is no evidence that Husserl
believed that transcendental egos could survive independently of
their being selves in the world. Furthermore, he believed that, though
people can live in radically different cultures, we all live in one and
the same nature. He regarded it as unimaginable that different
subjects should not recognise the same natural world. He is opposed,
thereby, to those who argue that nature itself is a 'cultural construct',
with the possible implication that there is no such thing as nature
itself to be respected for its own sake.

However, it is once again in Heidegger and especially Merleau-Ponty
that the most environmentally significant notions of the subject are
developed. Heidegger's name for human subjects is *Dasein*, literally

'being there'. To echo the way in which he puts it himself, one might say that he thought it was distinctive of the kind of being we have (or beings we are) that we are where we are (and as we are) in virtue of existing in a particular place. This is not to be construed geometrically but rather as involving being as we might say 'in place' in our environments. Our ability to interact with our environment is a more fundamental contact with it than our ability to think, conceptualise and theorise about it. Our ability to dwell in the world as our home is more fundamental, deeper, than our capacity to pass through places as tourists or commuters. Heidegger's emphasis is on practical rather than theoretical aspects of life. We engage with the world, and this engagement can be revealed, by phenomenological description, as the basis of all that we do. It can also be revealed to be, at heart, a caring engagement, though many of our activities disguise this care, alienate our caring or engaged selves, ignore the fact that we are *Dasein*.

Merleau-Ponty's notion of the subject is arguably the most significant environmentally. Merleau-Ponty more explicitly than Heidegger divorces being a subject from being conscious. He locates subjectivity in the body, and the subject is first and foremost a 'body-subject'. The environmental significance of this is immediately obvious. It is our bodies that are the environmental problem: they consume and pollute, and suffer from the results.

In the last section we saw how phenomenology puts emphasis on action rather than thought in its account of the relationship between subjects and objects. Merleau-Ponty goes on to emphasise the role of the body in action. The orthodox account of action, which Merleau-Ponty rejects, is that an action involves a conscious subject directing its body to move in certain ways in accordance with the wishes, desires or decisions that are features of consciousness or mind, i.e. mental activities. Merleau-Ponty invites us to pay attention to those features of our actions that are habitual. Much of what we do in the world is or relies on the fact that we have bodily skills that operate without our needing to pay conscious attention to directing them. We drive our cars, negotiate obstacles, play musical instruments, use sporting equipment, perform all manner of complicated tasks, as it were, on 'automatic pilot'. Our bodies know how to do things. Intentionality, that feature of subjects taken as central by phenomenology, is, according to Merleau-Ponty, at root a feature of bodies: our bodies point, grasp, direct themselves towards or away from things. He argues that, without this bodily intentionality, we would not have intentionality of consciousness. Our consciousness of the world and of ourselves always has at its base, as phenomenological description reveals, bodily know-how.

Merleau-Ponty spends a great deal of time articulating and arguing for this position. He also devotes much space to detailed criticisms of the orthodox accounts, both philosophical and psychological, of action. It is perhaps sufficient here to report some of the more striking observations he makes about human activities in criticism of the orthodox view and in support of his own position.

Consider any complex skill that you possess: driving a car, playing a sport or a musical instrument, using a computer, cooking, writing. With some of these, it might help if you cast your mind back to when you were learning the skill, or rather the large numbers of skills involved. In the early stages of learning to play tennis, I certainly found it hard to conceive how all the required elements could possibly be done at the same time. If one got one's feet in the right place, one forgot to watch the ball; if one got one's grip right, the feet got tangled up. If one watched where one's opponent was going, one lurched to volley a ball on its way out of court. If one concentrated on the back swing, the follow-through went wrong, and anyway one missed the ball. Gradually, all these things come together. The feet learn to take care of themselves, one's eye learns to look, the grip takes care of itself. But all this can go badly wrong again if you turn conscious attention to each element. Learning a skill requires that the body takes care of itself. Anyone who plays a musical instrument knows about finger memory and how it can be upset by thinking about it. Many of the skills involved in our activities have to be performed unconsciously. Psychologists tend to think here of the unconscious mind. For Merleau-Ponty, these skills are unconscious or as he says 'pre-conscious', but bodily rather than mysteriously mental.

Second, Merleau-Ponty observes that such skills come into play only in a suitable environment. In the absence of a keyboard, one's fingers lose their memory. Miming is a different skill. This looks like a problem for any account of action in terms of consciousness directing bodily movement. Why should consciousness not be able to do that wherever the body is?

Third, as we saw earlier, habits in a familiar environment can become so entrenched that consciousness cannot countermand them. You know there is a power cut, but your hand still reaches for the light switch. If the action is directed by the mind, why can't the mind stop directing it?

It should be clear from these examples that, when Merleau-Ponty describes the body-subject, he is not appealing to neurophysiology. He is interested in our bodies as we can become aware of them by phenomenological reflection on our life-worlds. He is concerned with describing, not explaining. Phenomenological description of the body-subject requires no special neurophysiological knowledge.

Merleau-Ponty's emphasis, then, is on bodily habits and skills. His descriptions bring one to ponder on how dependent one is on one's surroundings. In very practical environmental terms, one might think about habits that are environmentally friendly and those that are not. The easiest way of being economical with one's immediate environment would be to acquire entrenched bodily habits of switching lights and taps off, sorting one's refuse for recycling, re-using envelopes, etc. But such habits will need an appropriate environment in which to develop and operate. One can arrange one's own home appropriately, but if the bottle bank is a car ride away, then that could be a problem.

Merleau-Ponty argues that our basic bodily activities depend on our environment, and that our mental activity all depends on our basic activities. Conceptual knowledge involves perceptual skills. Perceptual skills involve bodily skills, focusing the eyes, moving round the object. Objects have significance – they make sense to us – inviting classifications, an appreciative gaze or touch, or a responsive movement. If this is right, it seems that it could have great import for the environmental debate. By debate, one might convince people that a certain way of life or course of action is right, what they ought to do. But getting them to do it, if the environment is not amenable to their acquisition of appropriate skills, might be an entirely different matter. Adjust the environment and the habits might follow.

Another consequence of Merleau-Ponty's position for the environmental debate is that part of the debate which concerns the ascription of instrumental value to the world of objects. Merleau-Ponty offers a rich taxonomy of instrumentality. Some instruments, tennis racquets, walking sticks, become part of the act, an extension to one's body. Others such as keyboards are occasions for the act. Others are the raw materials that we act upon – the joiner's wood, the dressmaker's cloth. Others – such as a workshop or kitchen – form a context for our actions. And all these operate against a background of the rest of the world being out there, taken for granted. I am not currently paying attention (or wasn't before I began this sentence) to the trees, grass, sky outside my window; but I would notice if they disappeared. So, using our environment as instrument, as means to our ends, is a complicated affair. The complexities not only concern the different kinds of means objects can be to serve our ends. Merleau-Ponty's work also casts doubt on the sharp distinction between agent and instrument, means and ends and the conception of ends often implicit in the claim that the environment has value as means to our ends. Ends are presented as something we consciously have or decide upon prior to going into action, using the environment as a means. This model is one which Merleau-Ponty implicitly challenges. We can formulate ends only if we possess bodily skills, and those skills can develop only in a certain sort of context. So our having the ends we do is dependent on environment. If that is so, is it so clear that the value of the environment can be just as a means to our ends?

Overall, the environmental message might be that, since our bodies are the problem, perhaps if we reflected more on ourselves as bodies and what they do, we might be better placed to solve some of the problems.

There is one last, but important point which Heidegger and Merleau-Ponty both make in their own ways, about being a subject in the world. Heidegger believes that as subjects we are capable not only of acting in and on the world of objects, we are also capable of a more reflective appreciation of it. He believes that the world, as well as showing itself as equipment or as raw material for our use, can show itself to us as it is. We need to do some clearing away of the plans, theories, practices which operate to show the world as we want to see it. But having cleared that away, the world will show itself in its true light. Merleau-Ponty voices a similar thought. If we pause from our everyday concerns and classifications concerning the world, we can experience wonder, awe, respect for the world. It is there and it makes sense to us. Both Heidegger and Merleau-Ponty illustrate this point with reference to artists. Heidegger cites Van Gogh as capturing and conveying in his paintings the world showing itself. Merleau-Ponty cites Cézanne as bringing out, in his paintings, how the world is something to be wondered at (see Plate 5.1).

Plate 5.1 *Cézanne's* **Mont Saint-Victoire**

Source: This painting appears in *The Art Book*, London, 1994, Phaidon, p.91.
© the Barnes Foundation

Objects

Phenomenological description reveals a very different world of objects from those described by the natural sciences and science orientated philosophy such as Cartesian dualism. The world of objects is not a purely factual realm, a collection of separate individuals, standing in contingent, causal relationships and obeying mechanistic laws. Phenomenology reveals it as a world of significance and value. Objects stand in internal relations to each other. They form totalities in which the whole is more than the sum of its parts and the whole makes the parts what they are.

One sort of totality which Heidegger emphasises is an 'equipment totality'. Imagine a workshop: the hammer has the significance it has only because there are also nails and wood. The sawdust is a sign of wood and of the saw. Tools are for use on raw materials to be reshaped into artefacts. To describe the workshop in purely factual terms would leave these significances out of account. It would involve abstracting from the life-world, ignoring those features of it which make it meaningful and valuable. Heidegger distinguishes between two ways in which objects can appear. They can be ready-to-hand or present-at-hand. In equipmental totalities, they are ready-to-hand. They become present-at-hand, mere things, when they break down, when they lose their significance as an integral part of the totality. For Merleau-Ponty too, objects form significant complexes. He emphasises also that they can often be ambiguous or indeterminate depending on the context or contexts they are in.

Those who seek to apply phenomenology to the natural environment point out that this approach may be more suitable to the study of ecosystems. They, too, are significant totalities with parts relating symbiotically to each other, properly regarded holistically, not as collections of separable items. Whether ecological science does, can or should adopt a more holistic, more phenomenological, model is too large an issue for this chapter. Suffice it to say that the current models – the system of energy model and the system of information model – both seem to be, in essence, mechanistic models that construe nature in the light of machines, albeit more complex machines than Descartes knew about (see Chapter 7). The issue is whether one can explain in mechanistic terms such features of the world as the normal size and life-span of members of a species being dependent on their habitat, or that how an ecosystem will develop, spread, withdraw, increase or decrease in biodiversity, will depend both on what its constituent parts are and the greater whole of which it is a part. Phenomenology, it is argued, could be used as the basis for a more enlightened ecological science.

It is also claimed that phenomenological investigation of nature reveals value in nature that is not just value as an instrument for us to use as means to our ends. The complexities, harmonies, balances, robustness, fragility and integrity of ecosystems constitute a demand that we treat them with care, that we do not violate or manipulate them. Others might claim that these features of ecosystems at least constitute a practical warning: because of these features, it is extraordinarily difficult to predict the results of our interference with them – we upset them at our peril.

The world of objects in our life-worlds is revealed by phenomeno-logical description as our home, our dwelling, our habitat. Merleau-Ponty emphasised the role of bodily habits and skills in subjectivity. He also emphasised how different contexts invite different skills, or activate different habits. The world that we inhabit is significant to us at this very basic 'pre-conscious' level. It is neces-sary for our actions and so our lives to make sense. Not only action but perception depends on the world making sense to us. Mostly we feel orientated in our world, and it is intelligible to us, not intellectu-ally, but sensorily. Things look the right way up, not upside down. Textures that feel rough usually look rough. Sounds indicate the direction and identity of their source. We are not simply located spatially in the world, but situated in a meaningful environment.

Heidegger, too, emphasises the role of action in our being in the world. He explores the relation between dwelling and building, and concludes that they are, if not the same, at least intimately interre-lated. Our way of dwelling in the world essentially involves building. Heidegger recommends that we build in such a way that our build-ings reveal rather than disguise dwelling. Our buildings should allow nature and the significance it has as our home to show itself. Giving a clear account of this difference is a difficult task, but finding exam-ples of buildings that do and those that do not demonstrate dwelling is not so hard. Heidegger's example is that of a river bridge integral to the life of the community residing at both sides of it, a meeting place, a way for the river-banks to show themselves as possible dwelling places, etc. He contrasts this with a dam, part of a hydro-electric power plant. This has the significance not of nature as our dwelling place but of nature as a resource, a potential source of energy, to be stored up, and used up by us so that we can make it possible to live as if nature was not our home, to forget how to dwell.

The otherness of objects

For Husserl, the otherness of objects consisted in the fact that our experiences of them can never be total. From whatever perspective we view an object, there will always be other perspectives that we don't have and cannot have at the same time. Objects always outstrip our powers to perceive them, and they may always defeat the expecta-tions we have of them, or surprise us.

Heidegger and Merleau-Ponty both believed that there was more to the otherness of objects than this. Merleau-Ponty included in his phenomenological descriptions of objects in the world, their 'aseity' or 'there-ness'. It is part of seeing a physical object, not only that we could touch it if we so chose, but also that it looks real. Of course, we can be wrong. Something that looks real might turn out to be a trick of the light. But that is no grounds for ignoring the phenomenon of 'realness' or aseity. It happens, and in the main it does not deceive us, and it is an important part of our sense of being at home in the world and not, for example, feeling as if we were in a perpetual dream or drug-induced state.

He believed further that wonder was the appropriate state to direct towards the world once we appreciate this amazing combination it has of otherness from us and significance for us. Clearly, if this is our fundamental relation with nature, then it is not free of value. It is perhaps more like the point at which any valuing gets started. The practical import of this sort of valuing is perhaps that our first thoughts should be to leave nature alone, to respect it as other than us, and to consider very carefully before we stamp it with our mark, turn it into something constructed by us – an artefact.

In broad terms, Heidegger would agree with this. The point of disagreement between Merleau-Ponty and Heidegger emerges more in their different styles and concerns. Heidegger believed that nature could and would show itself as it really is, but only to those who allowed it to. Heidegger believed that modern life, science, technology, industry, commerce, tourism and economics entraps us so that we forget, lose sight of, or become alienated from both ourselves and our world, making us live 'inauthentically'. Again it is hard to state clearly what the difference is between authentic and inauthentic ways of living. It is certainly hard to articulate how exactly things can 'show themselves' or how we would recognise it if they did. But we can recognise, in ourselves and others, examples of inauthenticity, of alienation: tourists rushing from viewpoint to viewpoint to return home too exhausted to develop their film; commerce and politics requiring that fishermen throw their catch back to pollute the sea; butter mountains, grain stores and wine lakes being destroyed while people are starving. All these surely involve a failure to recognise things for what they are, to treat nature mindfully and with care.

To speak, as we do, of nature as having only instrumental value, as merely a resource to be exploited, or to regard any other value being attached to it as mere sentiment, phenomenologists would claim, is to forget two important features of the natural world. First, it presents itself to us as something other than an instrument, and that bending it to our purposes often involves an obvious violation of something we know to be precious. Second, instrumental value is a much more complex thing than is perhaps generally recognised. To use nature as a means to our end, we need to have ends. But the ends we can have depend on the skills and habits we have developed. We have developed these in a world of objects. To then regard these means and ends as separable is to distort the relation of instrument to user.

Phenomenology is likely to be challenged by both sides of the environmental debate. It will be claimed by those who regard nature as having only instrumental value that phenomenology has not provided adequate evidence of any other kind of value. By those who think that nature has intrinsic value, phenomenology will seem too anthropocentric, talking as it does of how objects present themselves to us.

Phenomenology has the same reply to both of these objections. We need to look at the phenomena that underpin both of these kinds of value. A purely instrumental mentality is incoherent because it depends on an over-simplistic division of ends and means, and too simple a view of what using things as instruments involves. Such a mentality also risks entrapping us. As to intrinsic value, phenomenology is seeking to describe how we come to recognise it. Without an account of that, it is unclear what practical consequences could possibly follow from its attribution to nature or anything else. If you don't know how to recognise it, how can you possibly respect or protect it?

Applications

Phenomenology has applications both in theory and in practice. What are the practical applications? On a personal level, phenomenology can be life-enhancing. It encourages one to dwell on one's experiences, good and bad, and tease out what it is about them that makes them significant. It helps us to make sense of our lives and our relations with our environments.

Heidegger invites us to reflect on ways in which we are alienated. Older readers may reflect on how we, as a culture, are becoming increasingly alienated, or at least separated from nature. Our expertise both in interpreting, and acting in, the world is very high in the world of artefacts and technology, but much less so in the world of nature. In Heidegger's terms, we are increasingly enframed by technology. We take for granted the technological imperative and the technological fix. The world and the care distinctive of proper dealings with it are out of sight behind the technological interface.

Second, phenomenology could be a benefit to science. It might make the scientific investigator or observer more receptive to what they are observing. This might lead to good observations in support of or in opposition to established theories. It might also lead to devising new theories and hypotheses. In particular, some central concepts in ecology – ecosystem, niche, habitat, symbiotic relations – might be explored in the light of phenomenological discussion of internal relations between objects.

Third and relatedly, phenomenology can make us aware of the complexity of ways in which our lives are intimately tied up with our environments. We might thereby come to recognise this feature of any living thing, other beings with their environments. Some animals, our pets, clearly dwell, and in much the same way that we do. Might we have a more enlightened view of all animals and even plants if we attended more to how they dwell in their habitats? They do not experience the phenomena as we do, but we might use some of the concepts drawn from phenomenology to apply to other creatures' relations with their environments. Even non-living things, natural or artificial, might invite us to apply the terminology that phenomenology invites: the cliff 'protects' the cove, the mountain peak 'dominates' the range; the church 'nestles' in the valley; the spires 'dream' over Oxford's waking products.

Such descriptions, it will be objected, are metaphorical and not literal truths about the world. But what is literally true of the world is precisely what is at issue. Appreciating the aptness of these metaphors involves noticing precisely those features of the world that make it our natural home, which enable us to dwell in it.

Theoretically, phenomenology has metaphysical and ethical implications. It rejects the orthodox view of science as a full explanation of how the world works, presenting it instead as a highly selective world-view that systematically excludes from consideration the ways in which the world presents itself to us as significant and valuable.

Environmental ethics, if it accepts the orthodox view of science, is then faced with the task of superimposing values onto this realm of 'pure' scientific facts. Some claim that nature has intrinsic value independent of our ability to recognise it. Others claim that nature has only instrumental value: its value lies in its ability to serve our ends. These two views might seem to exhaust the field: value is a feature of objects or it is created by subjects; value claims are either subjective or objective. Phenomenology rejects this dichotomy. Value, or rather valuing, is a feature of the ways in which we interact with the world. Articulating this is a hard task made possible only if we drop the orthodox picture of the world and values.

Conclusion

In summary, phenomenology has both practical and theoretical implications for our dealings with the environment. In practical terms, it can alert us to a crisis that concerns not only the quantity of our supplies but the quality of our lives. The phenomenological method of description, applied to how we relate to our environment, both natural and built, can itself result in a richer, deeper experience of that environment bringing us to know better and be more attuned to it and it to us. What we learn from such investigations might aid us in planning our towns, managing our countryside, protecting our resources and our wild places.

Theoretically, phenomenology offers a world-view that might be of more use in debates about the environment than the orthodox one it challenges with its sharp distinctions between humans and the world, and between values and facts. Phenomenology offers us care as characterising our fundamental way of being in the world, largely covered up by modern life; ourselves as bodies situated in the world, with the world itself as significant and meaningful.

Phenomenology has been applied to morality with respect to humans. Dreyfus and Dreyfus[1] have developed a notion of moral expertise that involves a developed sensitivity to human situations and a propensity to act appropriately. It has been an implicit theme of this chapter that a comparable expertise with respect to the environment is a human capability that we should seek to explore in theory and develop in practice. This would be akin to an Aristotelian virtue, and, as such, an essential part of the good life for humans.

Questions

1 What is phenomenology's most promising contribution to addressing environmental problems?
2 Is there an alternative to the relationship between a person and the world around them that is supposed by science?

Further reading

Evernden, Neil, *The Natural Alien: Humankind and Environment*, Toronto, 1985, University of Toronto Press.

Hammond, M., Howarth, J. and Keat, R., *Understanding Phenomenology*, Oxford, 1991, Blackwell.

Husserl, Edmund, *The Crisis of European Sciences and Transcendental Phenomenology*, Evanston, IL, 1970, Northwestern University Press.

Krell, D.F. (ed.), *Martin Heidegger: Basic Writings*, London, 1978, Routledge & Kegan Paul.

Merleau-Ponty, M., *Phenomenology of Perception*, London, 1962, Routledge & Kegan Paul.

6 Coping with individualism

I explained in Chapter 3 how Romanticism developed a critique of the scientific dimension of the Enlightenment thought, and its reductionist, mechanistic, deterministic outlook. One of the alternative ideas put forward by Romanticism was that we should regard ourselves as *part of* nature rather than separate and 'relating to' it, a point taken up and developed in the present day, particularly by the Deep Ecology movement. In this respect both critiques of Enlightenment thought, Romantic and contemporary, include a rejection of 'individualism'.

'Individualism' is quite a dangerous term, in the sense that people use it in different ways, often in a quite subtle way. Arguments can get their plausibility from its ambiguity.

One sense is this: that the individual is the 'unit of analysis' of human life, rather in the way that the atom or 'corpuscle' was the unit of analysis when physical things began to be seen through the eyes of Modern science.

It is difficult to keep this idea in focus. It is the view that somehow whatever may be true of the groups that human beings belong to – communities, societies, nations, tribes, eras, classes, or whatever – those features are a consequence of what is true about the individuals that make the group up.

It isn't the view that human beings are by nature selfish. And it isn't the view that they are by nature anti-social. These two views of the human being are possible within individualism but so are the opposites (that human beings are social creatures, always on the look out for what they can do to help others). All these views are consistent with taking the individual to be the unit of analysis.

What then is the alternative? If individualism in this sense is the thesis that the group is constituted by the individuals that make it up, the reverse would appear to be that the human being is constituted by the group to which he or she belongs. Marx stated that: ' The real nature of man [sic] is the totality of social relations.'[1]

Environmental thought – radical environmental thought – finds itself constrained by individualism not because it wants to say that the human being is somehow a human being in virtue of being a member of his or her society, but – can I put it like this? – because it wants to express a parallel thought about the human being and nature. This is that the individual is what he or she is in virtue of their belonging to nature.

The rise of individualism

The rise of individualism was one of the great changes that marked the emergence of the Modern world, and it may help to look back and see how the human being – a 'non-individual' – was referred to before that shift.

Think of a block of flats. We may on occasion refer to the second floor, or the fourth window from the left, three floors up looking at the front. Here we are identifying bits of the building by reference to other bits. Supposing the window we are referring to here needed replacement. Once the job was done, would we have to refer to it in a different way? There would be a sense in which there was a different window there, but also a sense in which we would pick out the new window with exactly the phrase we had used before – fourth window from the left, three floors up (looking at the front).

There are two individual pieces of glass in this story of course, one that to begin with is the window, which then gets broken, and another that becomes the window. They could be given serial numbers. But in speaking of 'the window' our focus is not on either of these, but on something that is defined by its position in the building – fourth from the left, three floors up (looking at the front).

This is too gross to be an analogy for the medieval conception of, or attitude towards, 'the individual' – but maybe it serves as a pointer. In feudal society, a person was somehow *defined* by his or her place in society. A man thought of himself, as Alasdair MacIntyre explains, 'in terms of a set of established descriptions by means of which he situates and identifies himself vis-à-vis other men'.[2] (And MacIntyre is thinking of men *and* women here.) People are thus being thought of as more like windows than panes of glass. When the question is asked, 'Who do you mean?', the feudal answer would take the form: the one three levels up and four from the left. But the reference would be not to a block of flats but to the complex edifice of feudal society.

Human beings in the Modern world are not defined by their relative location in social space. I am defined as a person who was born at such and such a time of such and such parents in such and such a place, not by the position I occupy in the social structure. When I die there will not (I hope) be any question of a 'replacement'. Somebody else may be got to do my job, or even play my role in the family, but it will be a way of slighting my memory to speak of these people as replacing me as a person.

When people are defined in terms of their relationship to others, by their place in a (social) structure, how they should behave is given by the position they occupy. That you are a person is not very significant, and no obligations or duties spring from this fact alone. Your duties come from the fact that you are a knight or a king. One great change associated with the birth of the Modern period was a transformation in the significance of the person simply as a person. Just as the natural world began to be seen as existing independently of its role in bringing messages and being useful to human beings, so human beings began to be seen as existing independently of whatever use they were to society.

One site of this change was religion. Revolutionaries – Luther, for example – began to insist that, as far as God and His judgement were concerned, people stood *as people* – as individuals, denuded of whatever belonged to them in virtue of their place in the community. And as the significance of the individual grew, in that and in other ways, the question of how one should behave – as an individual and not as a 'position' – recruited interest and concern. Thus was set the agenda for Modern moral reflection.

Box 6.1

MacIntyre on the emergence of the individual

[In the transformation that gave us the Modern world] we get a move from the well-developed simplicities of the morality of role-fulfilment, where we judge a man *as* farmer, *as* king, *as* father, to the point at which evaluation has become detached, both in the vocabulary and in practice, from roles, and we ask not what it is to be good at or for this or that role or skill, but just what it is to be 'a good man'; not what it is to do one's duty as clergyman or landowner, but as 'a man'. The notion of norms for man emerges as the natural sequel to this process, and opens new possibilities and new dangers.

Thus the individual no longer finds his evaluative commitments made for him, in part at least, by simply answering the question of his own social identity. His identity now is only that of the bearer of a given name who answers as a matter of contingent fact to certain descriptions (red haired or blue eyed, labourer or merchant), and he has to make his own choice among the competing possibilities. From the facts of his situation as he is able to describe them in his new social vocabulary nothing at all follows about what he ought to do. Everything comes to depend on his own individual choice.

(Alasdair MacIntyre, *A Short History of Ethics*, London, 1966, Routledge & Kegan Paul, pp. 94–5 and 126)

Ethical thought in the Modern period has thus been pursued very largely within the individualistic framework. It has accepted the individual as the unit of analysis. In political thought there has been more attempt to explore alternatives. (Romanticism, as I have explained, saw a determined reaction to the science and scientism of Modernity, and actually reinforced individualism.)

In the remainder of this chapter I indicate two lines of thought within the Modern period with which environmental thinkers feel they can do business, as it were: the idea that morality is rooted in human feeling, developed by David Hume (1711–76); and utilitarianism, launched by Jeremy Bentham (1748–1832).

Environmentalists have proposed that utilitarianism may help us behave better towards the natural world if its principle, that we should act so as to maximise happiness, is not applied arbitrarily – e.g. to presently existing human beings only. If it were applied to all sentient beings, and to generations yet to come, it would move our behaviour strongly in the right direction. Others have argued for a development of Hume's approach instead: if moral behaviour is a matter of acting on *sympathy* what we need to do is to widen our sympathy so as to embrace much more of the natural world.

I then present the main ways in which individualism has shown itself in political thought – and action. Again, it is possible to attempt to amend the policy directives that would flow from a strictly individualist political philosophy to take account of some difficulties that appear from the environmental point of view, but radical environmentalists will look for something more fundamental.

Enlightenment moral philosophy

One way in which the scientific outlook shows up in the Enlightenment's tradition of ethical reflection is the way in which that tradition has avoided prescription. It has not given priority to the question of how we should live or what we ought to do. Instead its leading questions have been quasi-scientific. It has often taken as 'data' the judgements people as a matter of fact actually make about what is good and what is bad behaviour and tried to 'give an account' of that. One way of putting this project is to say that people have tried to work out what human beings and human thought must be like if these judgements of theirs are to make sense, to themselves or anyone else. What sort of creatures are they that can recognise a distinction between right and wrong, and apply it as they do?

Kant

The most sobering name to remember as an exemplar of this conception of ethical thought is Immanuel Kant's. Kant's thinking was in a sense quite revolutionary, but not in the sense that it informed a revolution in our conception of how we ought to live. He simply took the form of life instilled into him during a rigorously pietist childhood and pursued the question of what human beings must be like, and what thinking must be like, if the precepts he had so thoroughly absorbed were to be the inviolable and indubitable truths he took them to be.

Of course, some of the practical precepts of pietism come through into Kant's study of what makes moral judgements in general possible: for example the idea that what matters is that you should act out of a sense of duty, that you should never lie, whatever disaster your principle might bring in its train, or that you should avoid masturbation. These are judgements he thinks correct, but what he addresses is not the question of what makes them correct but what human beings and their thinking must be like in order for them to be intelligible as moral precepts. Kant's conception of the project of philosophical ethics was – and is – very influential. Much of contemporary and recent philosophical ethics is concerned to challenge and reform nothing – except 'the account we give' of moral *concepts* and moral *discourse*.

Even so, is there any inspiration we can derive from Kant? He had been inculcated with the belief that upright living was a matter of following principles 'religiously' – i.e. without allowing oneself any exceptions. What does a human being have to be like in order to live in this way? Kant's answer was that he or she would have to be subject to *reason*.

He thought that it was reason that made moral judgement possible, and that we exercised it by asking, 'Will I be consistent in doing it?', of anything that we thought of doing. One way of being inconsistent would be by doing something ourselves that we would disapprove of in others. Pouring greenhouse gases into the atmosphere while condemning others for doing the same thing would be inconsistent in this sense.

We can perhaps understand the idea that making exceptions to principles is a sort of inconsistency. We may also find plausible the idea that what vetoes inconsistency is reason, so that Kant's general approach may seem in a way promising.

But if we are asking not what makes the Ten Commandments view of the moral life possible but *how we should live*, the idea that we should act in accordance with reason may strike us as unobjectionable but also unhelpful. Whatever the circumstances, the test of consistency is likely to leave us with a multitude of possibilities still open.

I now turn to the first of the two promised lines of ethical thought. Both spring from the quasi-mechanical theory of the mind.

Moral behaviour is behaviour that is driven by feeling

Although few of the early Modern writers conceived of the human being as purely mechanical, like complicated clocks, they *did* think of them as *quasi*-clocks. That is, though they saw them as involving minds as well as bodies, they thought the question arose: how does the mind *work*? And they sought to answer this question by thinking of the mind having basic components (ideas) just as the physical world was thought of as having basic components (atoms), and by trying to articulate the quasi-physical laws to which the components were subject. Their leading question as far as morality was concerned was this: how to account for moral behaviour in terms of the quasi-mechanisms of the mind?

A basic conception of the quasi-mechanical view of the human being was put, for example, by David Hume. This was that 'the chief spring or actuating principle of the human mind is pleasure or pain'.[3] The early writers held that by linking pleasure to things that God wishes us to do (maybe for our own sakes – our well-being, or survival) He brings it about that we do them. By linking pain to things that God wishes us not to do He brings it about that we avoid them. For example what stops me from leaning on the stove and burning myself (on this theory) is that I feel *pain* before much damage has been done, and my feeling of pain brings it about that I stop leaning forthwith. (This general conception of the role of feelings of pleasure and feelings of pain in the government of human (and not just human, of course) behaviour is easily recast in a secular form: if not God then *evolution* has built the capacity for these feelings into us, for the same functional reasons.)

David Hume's theory begins with this: we act as the prospect of plea-sure and pain guides us, doing what promises to give us pleasure, and eschewing that which stands to cause us grief.

But then if this is so, how does it sometimes come about that I act to relieve the pain of *others*? It comes about because of the capacity I have for *sympathy*. When I encounter the signs of suffering in another, the ideal form of that distress transmutes inside my mind into a feeling of the kind whose effects I am observing. That is to say, I myself feel the distress. Sympathy involves you feeling what another person is feeling.

But because this distress that I feel is a form of pain, the prospect of such a feeling will lead me to attempt to avoid it. That is, I will seek to help others to avoid getting into distressing situations.

We have also, says Hume, come to set up some rules of conduct that seem to run contrary to self-interest. Respect for justice falls under this rubric. Sometimes, a human being will do something because it is 'just' even though it appears to run counter to his or her own personal interest. Hume accounts for this on the supposition that in a complex society it is sometimes in an individual's *long-term* interest to follow rules that in the short term appear purely altruistic.

Hume's attempt to show how morality is a function of the psycholog-ical make-up we are born with is echoed by modern thinkers, writing (unlike Hume of course) after Darwin and the acceptance of evolu-tionary theory. The evolutionary biologists asks: what is the function of 'morality'? How does moral behaviour, or the institution of 'morality', help the individual – or perhaps the species – towards survival?

Hume's thesis that our motivation for acting in others' interests lies in the *sympathies* that are built into human nature has also been taken up by modern environmentalists. If sympathy is what prompts us to act in the interest of others what we must do is somehow learn to *extend* our sympathies so that they embrace not only human beings but also animals, plants and ecosystems.

Utilitarianism

Originating within the eighteenth-century framework too, and from much the same perspective as Hume's, is the approach to morality known as 'utilitarianism'. Utilitarianism is often identified as the thesis that the guiding principle in morality is that you should act so as to produce 'the greatest amount of happiness for the greatest number of people'.

The starting point for utilitarianism was the same as Hume's: what motivated human beings, it was agreed, was the avoidance of pain and the pursuit of pleasure. But the perspective of the pioneers of utilitarianism, Jeremy Bentham above all, was different from Hume's. Bentham was interested in improving social arrangements rather than in accounting for human behaviour.

I have said that for the framework within which Bentham worked, the fundamental entity of the social world was the individual. And I have also explained that it was generally held in the early Enlightenment that what drove individuals – what brought about all their move-ments – was the pursuit of pleasure and the avoidance of pain. In the light of these two considerations if you then asked how society should be organised, you naturally fell into utilitarianism. That is to say, in the light of these two considerations, once you raised the ques-tion, it would seem obvious that you should organise things so as to make it possible for individuals to secure as much pleasure and avoid as much suffering as possible. Utilitarianism as an approach to social policy, as it were, is an expression of individualism.

Utilitarianism starts then as a principle of individualist social policy. But when it is taken up in the nineteenth century, principally by John Stuart Mill (1806–73), it is put forward as a rule to be followed by the individual person as they make their everyday decisions.
Utilitarianism thus becomes the principle that *each of us* ought to act so as to minimise suffering and maximise the amount of pleasure in the world.

Many people today seem to find this idea highly compelling. We don't always ask what grounds there are for accepting it, but are often moved by it, and moved to accept it as our own, without further ado.

From the individualist point of view, however, the utilitarian prin-ciple, that you as an individual have an obligation to maximise

everybody's pleasure seems in need of some justification. Why should an individual be concerned to increase other people's pleasure? One explanation that would satisfy an individualist of a certain sort would be that it was part of human nature to want to maximise the pleasure of others – that a human being had benevolence, as it were, built in. Another would be that human beings were such that they got pleasure from causing pleasure to others. It is the second alternative that John Stuart Mill in fact adopted.

There is another twist that Mill gave utilitarianism.

Bentham had couched it in terms of amounts of *pleasure*. Mill couldn't agree that all pleasures were alike. He wanted a version of utilitarianism which recognised that some pleasures were worth more than others – the pleasure of poetry for example more than the pleasure of winning at cards. He put this in the following way. Take a properly educated and mature person and observe what he or she prefers. *That* will give you the more worthwhile pleasures.

But this is a significant difference. The new principle is that you should maximise the extent to which people get what they *prefer*.

'What's the difference?', you may say. Pleasure is produced by letting a person get what they 'prefer', so maximising 'preference-satisfaction' is just another way of maximising pleasure.

But is it? Isn't it true that sometimes we appear to have a 'preference' of ours met but no pleasure ensues? If this does happen, we could describe the situation in either of two ways. We could say, though we *appeared* to have the 'preference', it turned out that we didn't really, or we could say that we had the 'preference' all right but in this case satisfying it didn't generate any pleasure.

Those who take the former option are recognising that a person can sometimes be mistaken about what they want. They may well go on to claim that we can be given a false conception of our own wants by propaganda, or by advertising, or through our education. A charge against the sort of political systems we are familiar with in the contemporary West is that they inculcate in us a set of 'false' wants, which the system is good at supplying. This then keeps the people from serious dissent, because it gives them a kind of pseudo-satisfaction. Dissatisfied, they nevertheless see their apparent wants being satisfied. They remain unfulfilled, but lack the clarity to analyse the situation and overturn the system that is keeping them in this state.[4]

Overt decision-making in contemporary 'democracies' relies heavily on the idea that a person's welfare depends on the satisfaction of the wants they think they have. Critics say this becomes completely indefensible, even as the best of a bad set of options, when the vested interests of those who have created the false wants proceed to meet them, and to grow fat.

It also plays a key role in 'cost–benefit analysis', which is the technique developed over the last three decades in an attempt to bring rationality to environmentally sensitive decisions. The 'benefit' of a proposed development is measured in terms of the degree to which it promises to fulfil people's wants – but it invokes the wants people *think* they have. If it is as false as critics say to think that what people think they want is a good indication of what they really want then the technique is broken-backed. Critics say that people think they want such and such a thing because they believe it to have value. We should be getting at that – the value that makes a thing desirable. Whether or not this is actually appreciated by people in any particular case is irrelevant.

Utilitarianism and environmental values

The jolt that utilitarianism has taken from recent thought comes specifically from the campaign for the better treatment of animals. The argument is simply as follows. If the principle is to maximise happiness, why should the calculation involve *human* happiness only? What of the happiness of other creatures who are capable of it? In other words, it is argued that those who believe in the principle of acting so as to make for the greatest happiness of the greatest number should not make any *arbitrary* restrictions in their application of it. There is no reason to think only of human beings: other creatures who are capable of suffering should in consistency have their likely experiences taken into account. This of course can alter quite radically the justifiability (in utilitarian terms) of a number of the things people do. The great bulk of modern Western animal farming methods would be completely unacceptable, for example, and so would the great bulk of animal experimentation.

It is by an appeal to this argument that some of the most effective environmental campaigning has been conducted in recent years. Its impact shows the grip that the utilitarian principle exerts. And this in turn suggests that the individualism presupposed by utilitarianism runs deep.

I have said that utilitarianism as an approach to social policy is little more than an expression of individualism. If individuals are all there are – if they are the 'units of analysis' – then, in so far as you have to arrange for them to live together, you can only have as your aim an arrangement that will let them get as much benefit as possible. This is an expression of individualism we ought to consider in a little more detail.

The liberal conception of the State

Individualism cannot simply insist that because there is nothing but individuals to think of, no thought need be given to social arrangements. This is because it is of course possible for one individual's activities to interfere with what another wants to do. Thomas Hobbes (1588–1679) argued that, human nature being what it was, mutual interference would take place. Indeed it would, he thought, take place on such a scale that no one would be able to do anything that they wanted! They would spend all of their time distracted from what they wanted to do by the need to defend themselves against a constant barrage of attacks. He argued that among individuals who were roughly equal in strength, a free-for-all would result in nothing-for-anybody.[5]

Hobbes, one of the pioneers of the Modern world, took a 'low' view of human nature. We are, he thought, by nature completely selfish: 'of the voluntary acts of men everywhere, the object is some good to himself' (women too of course), and it was their selfishness or 'egoism' that pointed to the need for a central authority.

Others have thought human nature more mixed, or even thoroughly altruistic. Individualists who take this view might conclude that the best arrangement would be for individuals to be left without interference from any 'central' authority. Left alone, they will have a concern for each other's welfare that will allow everyone to flourish. Individualists who think along these lines are espousing a kind of *anarchism*.

But even if – as an individualist – you think a framework is necessary (because otherwise, left entirely to their own devices, individuals would be likely to cut across each other), the object of the framework will be to allow people, as far as possible, to do what they want. They are unlikely to be able to do *exactly* or *fully* what they want once the framework is in place, but in a sort of trade-off they will be giving up some things they might like to do in order to secure a space in which to do others without the threat of interference. So the framework – or government – is there to enable the individual to do more of what he or she wants – more, compared with the poor level of freedom of action they would enjoy if no framework were in place. This is the liberal conception of government, or 'the State'. It is there to create as much space as possible for individuals to do what they want.

Earlier I raised the possibility of a person being *mistaken* about what he or she wants. If such mistakes are possible, however, there is of course a very black cloud hanging over the sort of framework individualism has sought to establish. Its aim, I have said, is to set up a framework that will maximise individuals' freedom to do 'what they want'. But this is only an obviously desirable aim if people are generally *right* about what they want! If I am allowed to do what I want, but I do sad things in the *mistaken* belief that they are what I truly want, the result is dissatisfaction and disappointment all round.

This is the source of concern that environmentalists, among other critics, have with the *market*, which in a certain form plays a key role in the framework we live under in the contemporary world. This is something I will explain in the next section.

The market

One part of the problem individualists are addressing when they consider what arrangements we need to put in place, if we are in any sense to live 'together', is how to make sure we don't cut across each other more than we need to – so I can do what I want, as often as possible, and so can you.

But there is another part. Living together allows individuals to pool their efforts and create for themselves an alternative to the life of absolute self-sufficiency. But if individuals are to co-operate in providing for themselves the necessities of life – food, shelter, clothing, warmth, etc. (and child care) – there arises the question of how this is to be organised. One solution would be for the *Government* to take on this extra responsibility. This would be to establish a framework of law to define and safeguard the citizen's freedom, but also to organise the productive effort, identifying the tasks that need doing and allocating them among the participating individuals. This would be a kind of State planning.

Another solution, if solution is the right word, would be, so some have argued, to do nothing! They have argued that left to themselves, within the minimal framework of the liberal state, citizens will voluntarily go in for a mix of activities that will meet everyone's needs – and meet them in a maximally efficient way. This is the market solution.

As an individual in the liberal state I will buy from others what I want, and in order to pay for it I will produce what others will be prepared to buy from me. In this way, everyone's needs will be met, and the mechanism producing this perfect state of things is entirely automatic. If only you set up a liberal state that creates sufficient freedom for people to do what they want, the satisfaction of everybody's needs will be secured.[6]

This is an introductory cartoon only, but I need to sketch it in order to present something of the impact of the lines of thought I have been presenting on contemporary issues. Few people today believe that if you leave individuals to do as they wish, within a framework of 'neutral' law and order, the market will ensure the establishment and maintenance of the sort of human life which would be generally admired. This is not because we have abandoned individualism but because most of those who are individualist, and who support the basic idea of using a market to organise our productive activity, nevertheless accept that it has flaws which need patching.

One of these flaws is that the market doesn't have any mechanism for protecting the environment from exploitation – at least for protecting those bits of the environment that don't belong to anybody. It is then open to everybody to use those bits as it suits them, and nobody pays. But in the end of course they are ruined, and no good to anybody.[7]

For those who support the basics of the market approach, a way has to be found somehow of ascribing a value 'artificially' to these important things that the market ignores, and to feed this into the buying and selling that the market orders. A highly influential method of doing so is to ascribe value to otherwise unvalued things in proportion to the strength of people's preferences. People might be asked to put their preferences in order of strength, or an ordering might be drawn up on the basis of observing their behaviour. The assumption I referred to above is then made, namely that people know what they want! It is assumed, that is to say, that people's true good will be increased in line with the satisfaction of their preferences. Money sums are then attached to reflect the strength of preferences expressed. Once again it is clear that this entire approach is compromised with the acknowledgement that people are often mistaken about their true wants.[8]

Growth and sustainability

Another flaw, it is claimed, of the market mechanism is that it drives towards ever more economic activity. Partly this is the result of the pressure it maintains for maximum profits to be made, but there is also in the background a political factor: if growth is not maintained, the inequality of wealth between different participants in the market will attract attention. There is a calculation, that is to say, within the political system that supports the market, and this is that the 'standard of living' of the average person must be allowed to go on rising, or else they may be more concerned with the way in which some people are much better off than others. This is a dangerous thought because, unchecked, it could lead to the abandonment of the market altogether as unfair.

There are influential voices now to be heard questioning, as an expression of environmentalism of course, the ideal of 'growth'. It seems almost a truism, they say, that unlimited growth of economic activity on a planet with finite resources cannot be sustained, arguing that the ideal of *growth* should be replaced with *sustainability*.

The principle of sustainability insists that the life we adopt ought to be in a kind of equilibrium with the things on which it depends. A form of life would not be defensible, it is argued, if it unnecessarily uses up any of the resources that are vital to it. For example, if human beings depend on water, they should be careful with it – so that with comprehensive recycling it lasted not just the life-time of those alive now, but for the indefinite future. Any given generation should leave behind them in the environment at least as much in the way of vital resources as was there for them.

Rights

I have said that in pre-Modern society a person was much more identified with their 'position' in the social edifice – with their 'place' as one might say. To this 'place' attached duties and obligations – for example, a landlord had to serve as a knight. But there were also understandings about what the landlord got in return – protection when working the land. With the emergence of the individual, the question arose: are there obligations and rights attaching to the individual *as such*? The classical theorists of individualism thought there *were*: 'all being equal and independent, no-one ought to harm another in his life, health, liberty, or possessions'.[9]

Two questions arise. What is it about individuals that gives them rights? And what rights *in particular* does it give them? It has been argued that what confers rights on individual human beings by the same token confers them on animals. There is the suggestion that the configuration of features that confers rights in both cases is 'being a subject of a life':

> Individuals are subjects-of-a-life if they have beliefs and desires;
> perception, memory, and a sense of the future, including their own
> future; an emotional life together with feelings of pleasure and pain;
> preference- and welfare-interests; the ability to initiate action in
> pursuit of their desires and goals; a psychophysical identity over time;
> and an individual welfare in the sense that their experiential life fares
> well or ill for them, logically independently of their being the object of
> any one else's interests.[10]

Individualism challenged

The large idea that radical environmentalism presents us with is that the individualism underlying all of the approaches just mentioned may need to be displaced. The human being is to be understood as part of nature – this is one kind of alternative. The human being is part of a community of living things – this is perhaps another. These suggestions will be taken a little further in Chapters 7 and 8. Here I shall point to the different kind of rejection of individualism that came out of the political thought that was part of Romanticism. It is the approach developed most powerfully by Georg Wilhelm Friedrich Hegel (1770–1831).

A society or nation at any particular time was conceived of by Hegel as the *expression* of something that could not be identified with any individual human being or any aggregate of individual human beings. It was the expression of the *Geist*. The history of a people was the progressive unfolding of its *Geist* – an unfolding that followed ('*dialectical*') laws of development that Hegel claimed he had identified. In the end this process was destined to flower in the ultimately ideal form of human life. Hegel characterised it as a life of rational freedom.

It is difficult for individualists to interpret Hegel's identification of something beyond the individual as anything other than the proposition that there is, besides the individual men and women who form society, another individual, namely a *Geist*. And the individualists pour scorn on it accordingly. But we may think of the *Geist* as Herder (Chapter 2) thought of a person's *self*. Not an extra ingredient, but that which exists as a potential from the beginning, that which is progressively expressed as development, personal or historical, proceeds.

This Romantic 'seed' also appears in the work of Marx, not so clearly in his conception of society as in what it is, for Marx, to be human. There is something in the human being that is fully realised or fulfilled in none of the forms of social organisation that have yet taken shape, Marx thinks. This something needs for its fulfilment that withering away of the oppression represented by the State, which will follow, according to Marx, the final collapse of capitalism. Irony seems too ordinary a word to describe the way in which Marx's ideas turned out to inspire regimes, spread across half the globe in the first half of the twentieth century, in which subordination of the human being to 'the State' was so ruthlessly uncompromising.

The goal of social arrangements under the individualistic conception of the human being is to allow maximum freedom for individuals to pursue whatever goals they might have. This invokes what has been called a 'negative' concept of freedom – the absence of constraint. But the 'seed' idea of the Romantics has been used to support a different concept. An individual is free in this new sense if he or she lives their life in development of the potential that is in them – if they 'fulfil' themselves. For thinkers like Hegel the prospect of such fulfilment is bound up with the state of development of the society they belong to. The fullest freedom is self-realisation of the individual within the context of a society that has fully realised itself.

Summary

- I have been describing one or two lines of argument that start off in the individualism of the early Enlightenment, but that come down to us, modified here and there, shaping our outlook today and offering the main options that the environmentalist has to work within or react against. Which of these alternatives is called for – amendment or replacement – is a puzzling question. The baffling ambivalence of conventional contemporary morality for the environmentally concerned is that it already condemns, or has it within it to condemn, most of what the environmentalist is condemning. People at large do not have to be persuaded that cruelty to animals is to be avoided if possible, or that genetic engineering calls for prudence, or that it is wrong to be so careless and profligate in the use of the earth that its integrity, and so the well-being of future generations, is put at risk. We generally accept these propositions, by and large. And yet our practices take little heed of them. Are we to say that conventional morality is in no need of reform, because it already yields the judgements environmentalists call for, or that it has failed us utterly, because of the devastation that is going on under its aegis?

Questions

1 What would be the disadvantages in finding an alternative to social life that is based on the individual?
2 When we make policy, should we consider the harm we may do to animals? On what basis?

Further reading

Callicott, J. Baird, *In Defence of the Land Ethic*, Albany, NY, 1987, State University of New York Press.

DesJardins, Joseph R., *Environmental Ethics*, Belmont, CA, 1993, Wadsworth.

MacIntyre, Alasdair, *A Short History of Ethics*, London, 1966, Routledge & Kegan Paul.

O'Neill, John, *Markets*, London, 1998, Routledge.

Regan, Tom, *The Case for Animal Rights*, Berkeley, CA, 1983, University of California Press.

Singer, Peter, *Animal Liberation*, 1975, revised edn New York, 1992, Avon.

7 Lines into the future

In Chapters 2–5 science was the target of environmental concern. It is science, either intrinsically or in the form it has actually taken, or the thought patterns that get expressed in it, that has been blamed for the state of the world. It is indeed difficult to argue that science has nothing to do with our environmental problems – since it is over-whelmingly involved in everything we nowadays do.

But it may be possible to find help towards solutions in science also. There is in the outlook of modern biology on human beings an insis-tence on our embeddedness in the natural world, which supports the idea that we ought to treat everything else with the same respect as (at our best!) we treat each other, or so it has been argued. And there is also in the science of ecology a way of thinking about networks of individual organisms that supports ideas of human beings belonging to, and having obligations towards, natural communities. Some also find support in ecology for the more radical thesis that it may be better to think of human beings not as individuals as such but as parts of some more inclusive whole.

The argument that a proper biologically informed view of the human being points unambiguously to the need for a more respectful atti-tude towards the natural world, to which we belong, is mounted with great skill by Paul Taylor in his book, *Respect for Nature*.

At our best, he argues, we in the West show respect for other human beings. But there respect stops short. Generally, we seem to behave as though human beings can do as they like as far as the rest of nature is concerned.

Sometimes we make an exception for *suffering*: we sometimes agree that we shouldn't hurt creatures unnecessarily. But, apart from that, it seems widely agreed that we can do what we like. We need to consider the effects of our actions on other human beings – on our descendants as well as our contemporaries – and we mustn't cause unnecessary suffering. However, there is nothing further to weigh in the balance when we are deciding what we might do.

Take for example a proposal to cut down some rainforest and install grazing meadows for cows in its place. A developer might be forced to acknowledge the need to compensate the human beings who are to be displaced, and she or he might have to avoid burning the animals alive, but no account whatever would be taken of how the destruction looks from the viewpoint of the *trees* or, say, the *waterholes*. It is true that a full consideration of human interests would bring in a wide range of different points. The interests of people in generations to come, wanting to visit the place as it is, would give us a reason for opposing the change, and human interests would dictate that we take account of the contribution the rainforest makes to the production of oxygen, and so on. But all this would be to look at the proposal from the viewpoint of its impact on human beings. What of the 'point of view' of the trees, the epiphytes, the streams, the different ecosystems that, woven together, make the forest up? In the conventional thinking of our culture, these 'viewpoints' seem to be ignored. And, it is said, this is what lies at the root of 'exploitation'. If we ignore a thing's 'point of view' then once we have consulted *human* interests we will feel free to do anything we like with it.

But we only 'ignore' the 'points of view' of plants and so on, it may be claimed, because there is *no such thing* as a plant's point of view. Plants, streams and the rest simply do not have 'points of view'. We are not wrong to ignore the viewpoints of these things because they have none.

A great deal appears to hang therefore on 'points of view'. What are they? Is it just human beings who have them? Or is it sentience that creates them, so that animals, or some animals, have them too?

Points of view

When one speaks of a point of view, the picture is of somebody perhaps on a cliff-top looking along the coastline first in one direction, then the other, then out to sea, scanning the horizon. Such an observer occupies a single position, and from this single position everything else is seen. The figure of the point of view suggests that *this* is what it is like for me as I go about the world, and go through life. I occupy a quasi-point from which I look out on everything that isn't me.[1]

This I think is a very familiar way of thinking about ourselves. It is perhaps saying little more than that we *are* 'selves'. 'I' am one thing, separate from the other things in the world, upon which 'I' 'look out'. This, as we shall see, is one particular way among others in which the 'self' may be conceived, but it is a familiar, conventional one.

Is it possible to imagine an alternative? Supposing I looked left and saw the cornfields, and right and saw the windmill, and then out to sea and saw the ferry *but didn't relate these experiences to each other*. Is that conceivable? Maybe I wouldn't in those circumstances be entitled to describe the situation as *me* seeing first one thing, then another, and then another. There would be a series of experiences, but if there was no linking between them the sense, surely, in which they were experiences of 'one person' would be lost. Would it still make sense to say that they were all experiences that belonged to one viewpoint? It wouldn't at any rate if we now imagined a person moving about instead of looking out from a single position on the cliff-top. Imagine, in *those* circumstances, the links between successive experiences being dropped. We then have a sequence of experiences, but nothing to connect them together, and nothing upon which to base the idea even that they are related to a single viewpoint. You may think, on reflection, that there is no real possibility here, that what I have suggested is strictly inconceivable. You may think, for example, that the idea of an experience is tied to the idea of a single subject having that experience as one among others: my attempt to get you to imagine an experience that is not linked to others and not thought of as 'mine' is conceptual sleight of hand.

The point at the moment is only this: the conventional conception of the 'self' is of one thing, separate from other things in the world, upon which 'I' look out. This makes a very close connection between having a point of view and being a subject.

But there is another dimension to the notion of a point of view that I need to bring out. You have a point of view on something if it *matters* to you. Unless you prefer some outcomes to others, I can't take your view into account as I decide what to do: 'can't' because once again there is nothing to take into account. If you lack preferences, all my options are equally attractive, equally unattractive to you. So if it is to play any part in altering our behaviour with regard to natural things the notion of a point of view has to be something more than the idea of a 'subject': a single thing looking out on a world made up of other things separate from it. It has to be something that values some situations above others.

One of the most obvious ways in which a situation might matter to you is if you are capable of suffering. Some situations may cause you suffering, and others may not. This amounts to saying that, for beings capable of suffering, some situations are preferable to others.

So here we have a basis for recognising a constraint on our actions. A creature that is capable of feeling pain, and perhaps pleasure, has a point of view that it is possible for us to take into account. There is no argument yet that we *have* to take it into account – just that here at any rate is something we *could* take into account if we chose to. There is a point of view here other than our own. A being looking out on the world made up of other things separate from it, and one that prefers some states of affairs to others because some things are painful and some things are less so.

A second way in which a situation might matter to you is if you have plans for the future that might be frustrated. Somebody's putting paid to your plans may not cause you physical pain, but it *would* be a case of a situation coming to pass other than one you preferred. So here is another case of a point of view that I *could* take into account if I wished to.

If animals are capable of suffering, they have a point of view in virtue of that. Our actions matter to them, in so far as they might occasion or alleviate pain. But what of other natural things – things to which the attribution of consciousness, and so of the capacity for suffering, is more problematic? The argument is that such things *may*, in certain cases, be regarded as having a point of view in virtue of the fact that there is a good sense in which they may be said to 'prefer' some possible futures to others. To take a strong example, think of a young plant. There is a sense, it is argued, in which a seedling 'aims' to grow into maturity, and so has a 'preference' that it is possible to take into account.

I have said that to have a point of view is to be a single thing, separate from other things in the world upon which 'I' look out, and 'preferring' some things – some present states, some future prospects – to others.

I have had to leave 'preferring' in quotes, however. What can be the basis for saying that a plant – or a stream, or a mountain – 'prefers' some possible futures to others?

I need to explore this a bit further.

A thing's 'nature'

One possible basis for saying that a thing like a plant might have a sort of 'preference' for some possible futures over others lies in the idea of a 'nature'.

Animals and plants, it is said, have *natures*, and their good lies in their being able to do what allows full expression to those natures – in being able to do what 'comes naturally' to them. A certain way of life comes naturally to a chicken, for example. (A way of life that is thwarted when the creature is tightly caged up in a battery.) It is their nature that gives them a 'preference' for some possible futures over others, and that represents something we could take into account in our own decisions if we wished to.

The recognition that animals and other living things have natures is a feature of many cultures – perhaps without exception – though there are different accounts of what a 'nature' *is*. In Chapter 9 I say something about the account that was given by thinkers of the late medieval period, who, drawing on Aristotle, identified the nature of a thing with what they called its 'form'. Today, however, conventional thought locates the nature of an organism in its chromosomes.

The metaphor with which we often think about the chromosomes and their role gives us the point that some take to be the key to the question of organisms having a good of their own. We say that the chromosomes act as the *blueprints*. This suggests immediately that from the earliest point there is in the organism a plan of the final outcome of its development, and thus also a kind of specification for its pattern of life. In the terms I used earlier, the blueprint gives the organism a 'preferred' future, both because it has a mature form to 'head for' and because its chromosomes equip it for a particular pattern of life that will be more suited to some possible futures than to others.

The chromosomes give an organism goals

In setting for the organism a form of life the chromosomes must also be setting for the organism a series of goals, organised in a hierarchy. For example, the pattern of life of a tortoise involves sexual reproduction. But if sexual reproduction is a high-level goal, the animal has to be equipped with a lower-level goal of achieving intercourse with an appropriate mate at an appropriate season (i.e. when there *are* appropriate mates to be had). But if intercourse is to be achieved, the animal must be equipped to pursue an even lower-level goal of locating such a mate. And if that is to be achieved by an animal fenced into a garden, for example, it might have to switch first into pursuit of an even lower-level goal of first finding a way out and into more promising territory.

Gilbert White wrote about the animals in his eighteenth-century Hampshire parish without knowledge of chromosomes, but his account of the adventures of a tortoise named Timothy records the goal-seeking behaviour that could be observed in him:

> there is a season of the year (usually the beginning of June) when his exertions are remarkable. He then walks on tiptoe, and is stirring by five in the morning; and, traversing the garden, examines every wicket and interstice in the fences, through which he will escape if possible; and often has eluded the care of the gardener, and wandered to some distant field. The motives that impel him to undertake these rambles seem to be of the amorous kind; his fancy then becomes intent on

> sexual attachments, which transport him beyond his usual gravity, and induce him to forget for a time his ordinary solemn deportment.[2]

In giving an organism a blueprint for its development the chromosomes give it one important goal, and in giving it a distinctive pattern of life they give it a whole series of others – an elaborate hierarchy of goals in fact, and the goals that the creature will be observed to pursue in its behaviour.

Goal-directed behaviour and consciousness

On the view of the organism I am sketching here, there is an assumption that a creature can perfectly well pursue goals without being conscious of them – without indeed being capable of consciousness at all. The view is that the workings of evolution can equip an organism with a control device – the brain – which on its own, and without any intervention of consciousness, generates behaviour that is directed at goals.

But is this possible? If we think of the brain as a control device governing behaviour can we think of the behaviour it produces as 'purposive' or 'goal-directed' at all? Isn't there a crucial difference between action that is engaged in because the agent considers alternatives, chooses one of them and pursues it, and behaviour that is produced by the essentially mechanical or quasi-mechanical – neurophysiological – processes in the brain?

This is a topical question in the field of artificial intelligence, where one aim is to get machines – electronic machines – to simulate different aspects of human intelligence. We can get machines to *appear* to engage in goal-directed behaviour in some settings. For example, an early wheeled robot was programmed to roll towards a light bulb placed in the centre of the room, and to do so even when obstacles of various sizes and shapes were placed in its path. More recently, the exploratory vehicle (the 'rover') that was so successful on the Mars mission of 1997 had been programmed so as to be able to negotiate the details of the terrain on its own while pursuing the overall destinations given to it, e.g. to move to such and such a place on the Martian surface.

Plate 7.1 *The Russian Marsokhod Rover displays 'goal-directed' behaviour that is somewhat animal-like, 1995*
Source: © NASA/Science Photo Unit

Are robots such as these truly pursuing goals? Or are they merely *simulating* goal-directed behaviour? To get machines *really* to pursue goals, do we have somehow to equip them with consciousness?

Another type of example to consider is provided by plants. Plants sometimes behave as though they have goals. If you move a pot plant about, for example, it will often repeatedly reorient its leaves in an apparent attempt to keep them exposed to the light. How are we to understand these movements? They seem to be purposeful – they seem to be carried out by the plant to maintain its exposure to light – but can we think of them as purposeful without thinking of the plant as conscious?

'Goal-directed' behaviour

We ought to pause to think through what it is about the behaviour of the clever robot, and the plant, that makes us think of them as apparently purposeful. In both cases, we notice in the first place, there is an end-point that the behaviour gets the system to – a geographical destination in the case of the rover, and an optimal exposure of the leaves to light in the case of the plant. But the second point is that the two systems are 'resourceful' in getting to that end-point. When a first tactic fails, they have the capacity to try something else, and if that should fail, then something else again. What they show, it has been said, is 'persistence towards some end state, under varying conditions'.[3]

This isn't a complete account of what is special about this kind of behaviour, though, because it seems to cover such things as water flowing downhill. In a reservoir full of water up in the mountains you have something that persists towards an end-point – sea-level – and something that has a good deal of capacity to find its way round obstacles. Let the water out of the reservoir and it will be tremendously resourceful, you could say, in finding a way down – a resourcefulness measured by the difficulty you would have in stopping it. It *might* be possible to do so, but the water would find a way round many of the steps you might at first think of taking. Would we want to say that water flowing downhill is apparently 'purposeful'? Would we want to say it appeared to be purposeful in the same way that the phototactic plant appears to be purposeful?

There are those who think that water flowing downhill and robots like the rover are not alike in their capabilities for this kind of behaviour. They propose adding a further condition to the formula just put forward.

Ends

They suggest that the crucial difference between a robot and water-flow is this. In the robot there is a representation of the end-state that has been given to the robot as its goal. But there is no such representation in the case of the water.

In the robot – the more straightforward ones, anyway – the goal is achieved schematically like this: there is in the program a stretch of code that represents the 'goal-state' to which the programmer wishes the robot to get to. At the same time, if and when the robot changes position, it is made to keep a note of its *actual* state. What the system is then programmed to do is to maintain a continual comparison of these two representations – the representation of the goal-state on the one hand and the representation of the system's momentary state. The outcome of this constantly updated comparison feeds into its choice of behaviour. Any behaviour that narrows the gap between actual state and goal-state will be persisted with. But, when a behaviour begins to have the opposite effect, it is discarded in favour of something else. In other words, we assume a system that has at its disposal a repertoire of behaviours. It might start things off by choosing one of these behaviours at random. But in its constant monitoring of its own position, and comparing where it is with the representation it carries of where it 'aims' to be, it can tell if that behaviour is narrowing the gap between actual state and goal. If it is bringing the two representations closer together, as it were, that behaviour will be persisted with. But if it isn't, the system will switch to something else in its behavioural repertoire. And so on.

Feedback

The special feature present here is called 'feedback'. The result of an action has an impact on what action is engaged in next. I have spoken of a robot doing x, and then checking whether x has got it any nearer its goal and then either repeating x or switching to y depending on the answer. In this way its mechanism incorporates a 'feedback loop'. I am incorporating a feedback loop in the control of my own behaviour if I check what my students have learnt at the end of a seminar and repeat my approach next time, or alter it, depending on the results.

It is suggested then that a distinct sort of behaviour can be distinguished, which I shall call goal-directed behaviour. This possesses the features of persistence towards an end state under varying conditions, which is achieved through a mechanism that incorporates feedback loops. It would be different from the behaviour of the billiard-ball, just sitting there until bumped into, and moving off 'blindly' in a calculable direction and at a certain pace. And it would be different from the behaviour of a body of water, cascading downhill under the influence of gravity. For with 'goal-directed' behaviour you have feedback loops, relying on there being some kind of representation of the end state – a stretch of programming, say, that 'codes' for the destination the programmer is designing the machine to move towards.

Purposive v. goal-directed behaviour

Programmable computers have made us familiar with the idea that a mere machine may be made to display the type of behaviour I am calling 'goal-directed'. But to say this, of course, is not to say that mere machines can be made to be *conscious*. When we get them to pursue goals, we are not relying on them having thoughts or plans. We assume that they are not conscious at all, but that, *nevertheless*, through the programming of behaviour control mechanisms that incorporate feedback they (or the robots they control) can be made to engage in goal-directed behaviour.

To try and keep our discussion clear, let us restrict the term 'purposive behaviour' to refer to the direct involvement of consciousness. 'Purposive behaviour', I will say, is behaviour that *does* flow from conscious thought on the part of the system doing the behaving. In those terms (assuming we don't think of the machinery as conscious) we shall have to say that computer-controlled robots can be made to display behaviour that *appears* to be purposive, but that isn't really. It is 'goal-directed' but not 'purposive'.

If we think back to Timothy (the tortoise, above), it may be true that as the creature appears to work its way round the garden fence we may be tempted to say it is searching for a way out. But, in the terms just set out, unless we are prepared to say that finding a way out is what Timothy has in mind, we cannot regard its behaviour as truly purposeful. It may be *like* behaviour that flows from a purpose, but unless the creature is consciously considering plans and choosing between them its behaviour is goal-directed but not purposeful. (There is a semi-jargon word, 'teleology'. Roughly speaking, any explanation that simply invokes the goal to which a particular bit of behaviour is directed, or the purpose for which it is engaged in, is a teleological explanation.[4])

I have been developing the idea that to have a point of view is: (1) to be a single thing, separate from other things in the world; (2) to look out on that world of other things; and (3) to have a kind of 'concern' over what happens, to have what I called in quotes 'preferences' for some future prospects over others. As I have explained, if a thing might be said to have a point of view is thought significant because acting morally involves taking into account all relevant points of view, and if a thing doesn't have a point of view it cannot figure in my ethical deliberations – I can't take account of something that is completely indifferent to all and any eventualities.

I have then tried to establish one sense in which a thing's 'nature' gives it a preference for some futures rather than others. If the nature of a thing is in part a system of goals, then there is a sense in which a future that sees those goals achieved will be 'preferred' by such a system to one in which they are thwarted. So having goals creates a sort of 'interest' in the future. I have tried to argue that this is independent of questions of consciousness. A system can have goals (though in the terms I used, not *purposes*) even though it does not enjoy consciousness. And so we reach the suggestion that the moral universe – the universe of things that have points of view, of things that are capable of being taken account of in ethical deliberation – is not restricted to conscious beings. Wherever you have something with a goal you have a citizen of the moral universe, an entity with an 'interest' in the future, and a 'point of view' that a moral agent *may* take into account.

Though I have introduced this notion of 'having a goal' via work with computer systems and robots (because it helps separate off the issue of consciousness), this feature of displaying goal-directed behaviour has often been taken to be characteristic of things that are *alive*.

Living things

Paul Taylor, the thinker who has been most powerful in developing this account of why we should show 'respect for nature', is amongst those that have taken this view. It is living things above all that are 'unified systems of goal-oriented activities'. It is living things, above all, that have 'points of view' in virtue of these activities – points of view that it is possible to recognise, and even to 'take'. With the right kind of extended and sympathetic study of an organism there comes a point when 'one is able to look at the world from its perspective'.[5]

Taylor's way of putting the point that a goal gives a thing a 'preference' for some possible futures over others is to say that such a thing has a 'good' – 'a good of its own'. In looking at the world from the point of view of another living thing:

we recognise objects and events occurring in its life as being benefi-
cent, malevolent, or indifferent. The first are occurrences which
increase its powers to preserve its existence and realise its good. The
second decrease or destroy those powers. The third have neither of
these effects on the entity.

Where a living thing is concerned therefore it is at least *possible* in our
decision-making to take into account 'what promotes or protects the
being's own good' as well as what is good for ourselves.[6]

Box 7.1

A good of one's own

Every organism, species population, and community of life has a
good of its own which moral agents can intentionally further or
damage by their actions. To say that an entity has a good of its
own is simply to say that, without reference to any other entity, it
can be benefited or harmed…We can think of the good of an individual non-human organism as consisting in the full development
of its biological powers. Its good is realised to the extent that it is
strong and healthy. It possesses whatever capacities it needs for
successfully coping with its environment and so preserving its existence throughout the various stages of the normal life-cycle of its
species.

(Paul Taylor, 'The ethics of respect for nature',
Environmental Ethics 3, 1981, p. 199)

Moral standing

Taylor is careful to say that none of the foregoing establishes that we
ought to be considerate where living things are concerned. All it
shows is that we could take viewpoints other than our own into
account if we wanted to. To put it another way: it shows that there
are viewpoints other than our own. There are things other than
ourselves, and other than sentient creatures, which have what I have
called 'preferences' as far as the future is concerned. These prefer-

ences derive not from the capacity of a creature to feel – pain or delight – but from the 'goal-oriented activities' that are characteristic of a living thing, sentient or not.

However, he goes on to argue that when we take into full account the fact that the human being is a product of evolution, and discard every consideration that is strictly irrelevant, we shall reach the positive view that living things of all kinds call on our respect.

Human beings belong to nature

To arrive at the conclusion that all living things demand our respect, Taylor develops three further claims, besides the one we have been considering, that living things have goods of their own. The first of the three is that all creatures are of equal intrinsic worth. I return to this in a moment.

His second is that human beings belong to nature on the same basis as other species. Living things, he observes, belong in general to a single large and complex network of interdependencies that comprises the earth's 'community of life'. The natural world, says Taylor, has to be seen as a unitary organic system. It is a system that maintains itself in dynamic equilibrium. Made up of many subsystems, each capable of change, it maintains a reasonably steady state by introducing compensatory changes as necessary. The subsystems are systems of the same kind – themselves made up of subsystems, and maintaining a steady state by introducing adjustments in some systems when changes in others would otherwise threaten it. What we have here is the contemporary picture of the living thing, applied to the biosphere as a whole (it is the Gaia hypothesis, under one interpretation). The part it plays in Taylor's ethical position is its claim that human beings are dependent on other species for their survival, and thus for everything.

Seriously, they are animals

The third of the further points Taylor puts in place is that human beings are *seriously* to be seen as animals, part of nature in the same sense in which other animals and plants are part of nature. Of course, almost everybody would take the view that the human being is an animal – whatever else she or he is. But what Taylor thinks we ought to accept is that 'our being an animal species' is 'a *fundamental* feature of existence'.[7]

These three further points, and the main one, together add up to what Taylor calls 'the biocentric outlook': living things have goods of their own, there is no sense in which human beings are superior to other species, human beings depend for their survival on the rest of nature, and human beings belong to nature on the same basis as other species. And these four points together provide the basis for *respecting* nature, and thus altering our comprehensively exploitative attitude towards it.

They do not, Taylor acknowledges, *prove* that nature ought to be respected. They do not represent premises from which the conclusion that nature ought to be respected follows by strict deduction. His claim is rather that, if you come to share the biocentric outlook, showing respect for nature will be entirely reasonable.

The biocentric outlook and respect for nature

This strategy of argument involves more subtlety than one would simple-mindedly wish (!), but it is not unfamiliar. If we take one particular element of Taylor's quartet, the claim that belonging to a biological species 'is an essential aspect of "the human condition"', we see facts adduced – the ecological dependence of human beings on other species, for example – but the role of the facts put forward is to do something other than prove a thesis. In fact, it might be said, it is wrong to think of the claim in question as a thesis at all. It is better regarded as a 'perspective'. The facts rehearsed in support have the role not of proving it as a conclusion but getting someone who reads them to shift their viewpoint.

In pursuing his strategy, Taylor sails close, but with skill, to what is called 'the naturalistic fallacy'. Facts, this observation has it, can never yield values: you can never deduce merely from how things *are* to how they *ought* to be. In the present case, the claim would be that no matter what the facts of biology might be, of human beings' dependence on other living things, for example, or of their common evolutionary origin, or of their common subjection to evolutionary mechanisms, nothing whatever follows about how human beings *ought* to behave. This observation is widely felt to be correct, and potentially devastating for the kind of argument Taylor appears to be making, with its apparent appeal to the scientific facts revealed by the scientific study of human beings and animals, and the ecological communities they form. Taylor is perfectly aware of this philosophical nuclear device, as it were, and he tries to evade it. He accepts that the facts he adduces do not *prove* that living things deserve respect – just that it would be only reasonable in the light of them to treat living things as possessing inherent value, and therefore, in the light of the other considerations adduced, as deserving respect.

He explains that the key step in the move from the scientific facts to the conclusion that we ought to respect nature is the denial of human superiority. He says that it is the scientific facts that 'result in' this denial, and, once we have made it, we are 'ready' to adopt the attitude of respect for nature.

This is a subtle, tricky movement of thought. It is best presented in Taylor's own words (see Box 7.2).

Box 7.2

The biocentric outlook grounds respect

Here, then, is the key to understanding how the attitude of respect is rooted in the biocentric outlook on nature. The basic connection is made through the denial of human superiority. Once we reject the claim that humans are superior either in merit or in worth to other living things, we are ready to adopt the attitude of respect. The denial of human superiority is itself the result of taking the perspective on nature built into the first three elements of the biocentric outlook.

Now the first three elements of the biocentric outlook, it seems clear, would be found acceptable to any rational and scientifically informed thinker who is fully 'open' to the reality of the lives of nonhuman organisms. Without denying our distinctively human characteristics, such a thinker can acknowledge the fundamental respects in which we are members of the Earth's community of life and in which the biological conditions necessary for realisation of our human values are inextricably linked with the whole system of nature. In addition, the conception of individual living things as teleological centres of life simply articulates how a scientifically informed thinker comes to understand them as the result of increasingly careful and detailed observations. Thus, the biocentric outlook recommends itself as an acceptable system of concepts and beliefs to anyone who is clear-minded, unbiased, and factually enlightened, and who has a developed capacity of reality awareness with regard to the lives of individual organisms. This, I submit, is as good a reason for making the moral commitment involved in adopting the attitude of respect for nature as any theory of environmental ethics could possibly have.

(Paul Taylor, 'The ethics of respect for nature',
Environmental Ethics 3, 1981, pp. 217–18)

The obligation to respect living things – to show respect for 'nature' – is what Taylor thinks is revealed by his analysis, 'respect' that consists in commitment to a way of life in which one seeks to promote and protect the good of all living things – for their own good, irrespective of any human interest. One will observe appropriate rules of conduct, and will seek to develop appropriate virtues. Broadly speaking, rules that give proper expression to an attitude of respect for nature will require us to harm living things as little as possible, allowing them to pursue their own lives, and to do what we can to repair such damage as is done to them, whether by ourselves or others.[8]

Value egalitarianism

Most striking of Taylor's four claims is that all living things are of equal 'intrinsic' worth, and I find it impossible to move on without raising an eyebrow.

Taylor acknowledges that from my viewpoint some living things have more value than others. My pet dog, for example, means more to me than a beetle on the other side of the world. But put *my* viewpoint on one side. And put everybody else's viewpoint on one side. What about the value of the dog, and the value of the beetle independently of anybody's viewpoint? What about what one might call the 'objective' value of these creatures? Taylor thinks he has shown where the objective value of a living thing comes from. He thinks he has shown that it is there in virtue of its goal-oriented activities. *This* is the basis, we have seen him argue, for saying that the living thing has a 'good'.

But then it would follow that a thing either has objective value or it doesn't – since either it exhibits goal-oriented behaviour or it doesn't. Objective value would then not admit of degrees, and comparisons between the beetle and my dog in respect of objective value would not make sense. The goods of different animals and plants would differ according to species, but there is nothing by which they may be ranked in order of priority. As living things they both pursue goals, and that confers the status of possessing objective value on both.

What is more, human beings are in the same boat. They are goal-pursuing creatures, and they would therefore be deemed to possess objective value: but no more than the dog, and no more than the beetle. Every living thing would have objective ('intrinsic') value, but none more than others.

These are conclusions that Taylor in fact embraces, though they have by no means always and everywhere been believed, he admits. The idea that inherent worth varies lies at the foundation of societies structured by class, where people of different 'rank' were certainly regarded as possessing different worth. But it is a baseless belief that – in the socio-political sphere, Taylor thinks – we have generally come to acknowledge. The idea that all people are equal – not equally good at everything, but equal in inherent worth – has become, he says, an unquestionable nostrum of the modern democracy.

Such egalitarianism however stops short at the human/non-human border. While all human beings are 'in modern democracies' considered to be of equal worth, the general sentiment is that human beings are worth more than other animals – indeed that other animals have no worth in themselves whatsoever. Taylor cites three sources of nourishment for this powerful belief. One is the contribution to Western culture of ancient Greek thought, which defined the human being as the *rational* animal. Human beings have part of their nature in common with other animals – but it is the *other* part, the reason, which 'enables us to live on a higher plane and endows us with nobility and worth that other creatures lack'.[9]

Cartesian dualism is a second source, with the idea that only human beings have souls: it is their possession of souls that makes human beings uniquely valuable. Taylor's comment is that even were some kind of dualism to be correct, there would still be no independent reason for thinking that having the power of thought makes one thing more valuable than another.

And the third is the Judeo-Christian notion of creation as a 'Great Chain of Being'. God created the world as a hierarchy, with angels and human beings towards the top and towards the bottom; beyond were the lower animals and plants, inanimate nature. Taylor sees no good reason for accepting the metaphysical framework that supports this picture. The Judeo-Christian tradition, he concludes, no more than either of the others, fails to leave us with any reason for discriminating against non-human living things.

Summary

- Taylor is drawing here on the modern scientific perspective on the nature of a living thing, and the nature of the human being. The living thing, a goal-directed system, has a line into the future that it is possible for human beings, in pursuing lines of their own, to cut across. There is something there for the human agent to respect, if he or she chooses to. And because of our place in nature alongside all other living things, we ought so to choose: that is, living things have a call on our respect.

- In the next chapter I turn to a line of thought that has much in common with Taylor's, though it comes from a different angle, and it aims to cast the net of obligatory concern wider. Where Taylor uses the biological notion of a living thing to locate the individual firmly within nature and firmly alongside other living things, the arguments we are to consider from ecology use the notion of a goal-directed system to argue that we belong to natural 'communities', within which we owe the obligations of membership.

Questions

1 What's wrong with filling in a pond?
2 What's special about life?

Further reading

Adams, Carol, *Neither Man Nor Beast*, New York, 1994, Continuum.

Clark, Stephen, *The Nature of the Beast*, Oxford, 1984, OUP.

Pluhar, Evelyn, *Beyond Prejudice*, Durham, NC, 1995, Duke University Press.

Taylor, Paul, *Respect for Nature*, Princeton, NJ, 1986, Princeton University Press.

Woodfield, Andrew, *Teleology*, Cambridge, 1976, CUP.

Wright, Larry, *Teleological Explanation*, Los Angeles, 1976, University of California Press.

8 Ecology and communities

In Chapter 7 I discussed how the conception of life sponsored by modern biology can be used to argue that living things as such demand our respect. In this chapter I now look at the second mine of environmentally helpful ideas that have been discovered by science – the ones associated with the science of ecology.

Ecology studies the ways in which individual animals and plants live in the natural state, exploring their interactions. It has established that their relationships are typically ones of mutual dependence. The most potent point environmentalists have taken from this is simply that human beings must belong to such systems of intricate mutual dependence too, and that our present way of living imperils us because this is what it ignores. You can also find in ecology an example in science of the idea I raised earlier (see Chapter 1) – individual organisms belonging to a larger whole might be thought of as sinking their identity in the identity of the whole itself, which is to be thought of as a kind of 'super' organism.

It will be easiest for me to introduce the ecological perspective by sketching key stages of its evolution.[1]

Ecology

It was the geographical perspective that first suggested the scientific significance of thinking in terms of *types* of vegetation rather than exclusively in terms of individual representatives of species. Why did grassland predominate in some regions of the earth, forest in others, marshland in still others, and so on? These were questions that posed themselves (to von Humboldt (1767–1835)) at the beginning of the nineteenth century. The point was then developed that you could have what should be regarded as the same 'vegetation type' – what were called 'formations' – even though the particular species present might vary. 'Tropical forest' came to be defined, for example, as a distinctive *pattern* of vegetation that might be made up of one set of actual species in Africa, another in South America and a third in India.[2]

The concept of a 'formation' carried the implication that some types – or types of types – of plants habitually lived together while others didn't. So you have established a minimal sense of plants forming 'societies'. What was then added, in a synthesis that launched the science of ecology proper at the end of the nineteenth century, was the proposition that the members of these natural societies *depended one on another* for their flourishing and indeed survival, making up 'communities' of plants and animals bearing 'manifold and complex relations' to each other.[3]

Several degrees of mutual 'support' were noted among the recognised 'communities' of living things. At one pole, species apparently 'co-operated' simply by specialising in different components of a single source of nourishment, so that though they 'ate off the same table', they took different things from it. At the other pole, the intimate degree of mutual dependency called 'symbiosis'[4] was recognised. The lichen, for example, which looks like a plant, turns out to be two plants of very different kinds, in close 'co-operation' with each other, a fungus and an alga.

The focus then shifted to the *changes* to which formations were subject. Changes could come as a result of external impacts, like the arrival of beavers in a stream, or the accidental burning of a forest, or they could come from within. Either way there would be created opportunities for species that had been alien to the formation. The establishment of these invasive species would in turn though create yet a different complex of conditions, which would create another set of opportunities, and a further wave of invasion would result. And so on.

The thesis emerged that, in any particular region, this series of inva-sions would sooner or later come to an end. And thus that for any particular region there was just one type of formation that was stable, and all vegetational change would in the end arrive at that point.

As the great pioneer of ecology F.E. Clements presented it, vegetation developed towards the climax form through a succession of discrete stages in a progress he called a 'sere', and a sere contained within itself a principle of change. That is to say, he maintained that each developmental stage as it became established began to create the conditions for its own replacement: for example by altering its own leaf fall the character of the soil was changed. Most vegetation, he apparently thought, was, at any one time, caught up in such a 'sere', so that nature in general could be characterised as a 'flow toward stability'.[5]

This emphasis on the regimes of change to which communities of plants were subject became central to the definition of 'ecology' as it gained scientific recognition. Ecology became identified as 'the science of the development of communities'.[6]

It is clear that Clements thought of a sere's development as 'goal-directed', in the sense explained in Chapter 7. It was quite possible he thought for the impact of unusual local conditions to 'divert' the flow of a sere, but (within limits) it would somehow have the resources to 'get back on track', resuming its progress towards the relevant climax formation. The question arises, of course, of what these 'resources' might be. Here you have the proposition that systems of vegetation pursue a goal. How exactly was this thought of as being achieved? Clements' answer is perhaps disappointing. He said it pursues its goal by the same mechanism by which an individual organism pursues a goal. In fact, a formation simply *is* an organism. We return to the issue here in Chapter 9.

The idea that there were a small number of zones on the surface of the earth, each defined by climate, that associated with each was just a single climax type of vegetation, and that vegetation everywhere within that zone was driven to develop towards the climax formation – this clear and powerful idea was challenged by proponents of what became known as the New Ecology, which took shape from 1925. The focus shifted from global patterns of development to the structure and organisation of particular communities. The concept of the *niche* was formulated, defined as the 'status' or 'occupation' of a species within a community. The notion of an *ecosystem* took the

New Ecology further in the same direction. It drew on the concept of a *system*.

This notion of a 'system' was being developed in the 1930s as a means of laying stress on functional relationships. A very fruitful analogy was being explored by the movement known as 'cybernetics' between control of animal movements and control of complex machinery. Animal control systems were thought of as assemblies of nerves and muscles, whereas the electromechanical control systems of the time were constructed out of wire conductors and lumps of steel. But in spite of such contrasts in material composition it was claimed that the two could usefully be looked at in parallel. If features to do with physical structure and composition were disregarded and emphasis placed instead on what *work* the various components performed, on the one hand in the animal and on the other in the machine, the two could be seen as addressing the same problem.[7] The point of calling a thing a 'system' was therefore to insist that it could be regarded as a complex with parts, each of which contributed to the functioning of the whole.

An animal has not always been regarded as a complex with parts contributing to the functioning of the whole, but that picture is familiar enough today. It means there is a question besides: What is its structure? We can also ask: How does it work? It means that as well as analysing an animal at the level of *physical* parts it also makes sense to analyse it as *functional* subsystems. From one point of view the lung is a lump of a particular tissue in a particular kind of shape. From another it is an organ that performs the function of oxygenating the blood.

In proposing the concept of an ecosystem, in the 1930s, botanist A.G. Tansley was therefore announcing his intention to treat what had been called 'communities' of living things *functionally*. He was going to regard them as machines – 'running' machines – with parts thought of as submachines contributing to the 'running' of the whole.[8] Tansley's ecosystem concept also brought the immediate inorganic environment into the reckoning. Soil, water, sunshine, temperature – all were treated as belonging not to the factory's environment but to the factory itself.

We have now assembled the leading ideas of ecology[9] as they are drawn on in discussions of the environment and our place in it. There are different ideas there, indeed ideas in conflict, but also a common theme, which is that the lives of individual animals and plants are often, perhaps always, entwined with one another.

The most radical idea is that the 'formation' is to be regarded as one organism rather than a succession of such – that what we ordinarily regard as a community of living things might in fact be a kind of super-organism. But not much new light is thrown on what this might amount to, except by saying that such a thing would have to be an organism, and thus alive. What I discuss now is the nourishment ecology gives to the idea that clusters of living things, to one or more of which we human beings may belong, might be regarded as 'communities'.

Communities of organisms

One of the seminal figures of the modern environmental movement is Aldo Leopold (1887–1948). During the first part of his career, Leopold worked as a forest manager in what was the conventional mode of the time. He worked within a framework which assumed that essentially forests were resources to be managed as yielding a cash crop. Leopold subsequently underwent a kind of conversion, coming to see that conventional attitudes were flawed and indeed dangerous. The fruit was *A Sand County Almanac*,[10] an eloquent plea for a revolution in our approach to nature, in which his stress is on the fact that as human beings we belong to a *community* of living things, and that our community is something we must nourish and support if we are to go on getting back from it the nourishment and support it provides in return. If we heedlessly exploit the natural things around us for short-term goals we will end up destroying the very support we need for our own well-being and indeed survival.

He offers in *A Sand County Almanac* a distinctive vision of our place in nature, drawing on the one hand on his familiarity with scientific ecology and on the other his profound feeling for 'nature'. The viewpoint he established, sometimes called 'ecocentric', has inspired others to develop in a more 'philosophical' mode the idea that ecosystems, such as forests and lakes – and not just human beings – must be treated as possessing moral significance in their own right.

Plate 8.1 *Grey Owl, an early environmentalist, sought to demonstrate a proper relationship with the wilderness in his life with a beaver colony in Saskatchewan in the 1930s.*

Source: Half-tone photograph, plate facing p. 225, in *Tales of an Empty Cabin* by Grey Owl, London, 1936, Lovat Dickson

The need for and possibility of the Land Ethic

Leopold took from ecology the idea of there being in the natural world, at a level higher than the individual plant or animal, 'complexes' or 'systems' – 'communities' – consisting of many individual plants and animals, of several different *types* of animals and plants, indeed, and also of a variety of other kinds of things as well – rocks and soil and water, for example. As we have seen, these complexes ecology has come to call 'ecosystems'.

It was borne in upon Leopold that in some circumstances ecosystems fall into decline – or, sometimes, pretty well collapse. It was deterioration of this kind that Leopold, as he wrote, saw transfiguring the countryside across large tracts of America in the 1940s.

The vast open grassland of the prairies, the heart of the North American land-mass, was turning to dust. In a process that went back no more than four or five decades, pioneering cultivation had used the plough on turf that had before then remained unturned since time immemorial. The soil, so long protected from the wind and the rain, was protected no longer. A series of minor alarms flowed from this, and then something entirely major: the great Dust Bowl catastrophe of the 1930s, when storms of unprecedented ferocity lifted the topsoil almost in its entirety and deposited it thinly, as dust, elsewhere.

Prairie grassland had been one of the strongest examples of a climax formation in the thinking of ecology's pioneers. This was so for Clements, for example, for him a paradigm of a fully mature and stable plant configuration. Stretching for thousands and thousands of acres across the centre of North America, the prairie had existed, so he thought, with much the same mix of species for thousands and thousands of years. That great span of continuity, transcending the time-scale of the human species, was broken in those few decades from the 1890s on by the invading White man, inventing the steel plough and using it to ribbon and upturn the turf.

The Dust Bowl that eventuated came to symbolise for many the prospect for human beings and their planet if they went on exploiting natural resources heedless of any limits. In the phrase made famous by someone who thought the whole world had been turned upside down, and not just the prairie turf: we could look forward to a *wasteland*. For those familiar with the ecology of the time, thinkers like Leopold, the diagnosis was that a finely balanced network of biological mechanisms – the grassland in its stable state – had been destroyed by a spanner hurled in by heedlessly exploitative agricultural techniques, of which the steel plough was simply the most damaging.

That was the point expressed in ecological terms – the outside observer noting the destruction of some of the finely adjusted mechanisms that are recognisable in nature. Leopold asked us to think a little differently however. We could not retain the status of the external observer in this matter, because our own fate was bound up with events that were in train. What we had to recognise, he urged, was that the 'systems' that were being destroyed were systems to which we ourselves belonged. In fact, the key system was one that embraced on the one hand what we call our environment and on the other, ourselves. Together, he urged, we belonged to a single system,

Plate 8.2 *'Prairie: do not disturb'*

Source: Colour photograph appearing in *National Geographic* 174(6), December 1988, p. 833. © *National Geographic*

which he wanted to regard as a single *community*. Ecological systems were not to be seen as all of them external to human beings, to be manipulated as human beings chose. Human life depended on other living things, was itself caught up in the complex web of interactions that maintained, but only within limits, a stable configuration. For this complex web Leopold invoked the resonant term 'the land'.

The land as a 'community'

What is the significance of Leopold's insistence on 'community'? First and foremost, to say that humans, and the animals, plants, rocks, soil and water they live amongst, belong to the same community is being used to express the dependency of human well-being on the well-being of the environment. It is then a valid and useful piece of rhetoric. It seems true that human beings depend on these other

---

Box 8.1

Leopold's Land Ethic

- 'The despoliation of land is not only inexpedient, but wrong.'
- A land ethic calls for willingness to perform 'an unprofitable act for the sake of the community'.
- 'We have no land ethic yet, but we have at least reached the point of admitting that birds should continue as a matter of biotic right, regardless of the presence or absence of economic advantage to us.'
- 'The "key-log" which must be moved to release the evolutionary process for an ethic is simply this: quit thinking about decent land-use as solely an economic problem. Examine each question in terms of what is ethically and aesthetically right as well as what is economically expedient. A thing is right when it tends to preserve the integrity, stability, and beauty of the biotic community. It is wrong when it tends otherwise.'
- 'In short, a land ethic changes the role of Homo sapiens from conqueror of the land community to plain citizen and member of it. It implies respect for his fellow-members, and also respect for the community as such.'
- 'But the education actually in progress makes no mention of obligations to land over and above those dictated by self-interest.'
- 'No important change in ethics was ever accomplished without an internal change in our intellectual emphasis, loyalties, affections, and convictions.'
- 'It is inconceivable to me that an ethical relation to land can exist without love, respect and admiration for land, and a high regard for its value. By value, I of course mean something far broader than mere economic value; I mean value in the philosophical sense.'

(Wye College External Programme, *Environmental Ethics*, 1995, University of London, Unit 9, p. 21)

---

things for their own good, and it also seems that this is a truth that requires reinforcement.

Emphasis on the community as an 'ecosystem' highlights the fact that relationships between natural things may be *subtler* than at first sight appear. There are more ways of destroying a community than ones that involve frontal destruction of its members, and some of these ways may be completely unknown. Small alterations may turn out to have unlooked-for effects.

At the same time the model of the community as ecosystem argues that it is characteristic of a community to show a degree of resilience in the face of potentially disruptive developments. A community pursues the goal of its own survival. The point is not therefore that a community is susceptible to every subtle change in its circumstances but that the impact of any particular change is often unclear and may be disastrous – may hit upon a system's Achilles' heel, a vulnerability no one knows about. Close the village post office when the person who ran it retires and you have not on the face of it destroyed any of the village's residents. But you may have destroyed 'the community' nonetheless.

Communities and their members have different time-scales

There are also ways in which a *human* community may be said to 'transcend' the lives of the individuals who belong to it – and here we begin to encounter, so some would say, the limitations of individualism. For example, in a human community ways of doing things are established (eating, growing things, looking after children, talking, including using a particular language) and are perpetuated irrespective of the individuals involved. Some of these are solidified as 'institutions', like, for example, councils of elders, or committees. Certain kinds of foundational beliefs are passed from one generation to another.

It is difficult to map all of these features of human communities onto the natural communities of which writers like Leopold are attempting a definition. But in traditional communities at any rate the pattern of human life undoubtedly interlocks with patterns in the life of the animals and plants, and the seas, lakes and forests that are their 'habitat'. This is perhaps enough to justify talk of a single community embracing living things in a given locale, the particular living things that are human beings, and their largely shared habitats.

And such a way of talking might be important because of the context it builds for human life. It invites us to see our life as a contribution to a pattern that is wider and bigger. What is important, it suggests, is that what I do gets done, but not that the particular individual that I am should do it. And perpetuating the pattern can come to seem more important than pursuing any particular projects that an individual as an individual may have.

The individual is moving away from centre stage here. It is the community that is becoming the 'unit of analysis', with the individual beginning to be redefined as a place-filler.[11]

The circle of our sympathy

A further note struck by 'community' is significant too. As we have said, in a human community, each member, by and large, depends on the well-being of others. You might put this by saying that this makes it in an individual's interest to think of others.

But there is something on top of this. In a community, individuals are typically concerned for each other's welfare *irrespective* of any impact it may have on their own. If a disease is infectious, I will want a person suffering from it cured as quickly as possible – because, for one good reason, this will minimise my own chances of catching it. But even if this kind of self-interested reason doesn't apply, I may still want to do what I can to help the person get better. A way of expressing this kind of reason is to say that the sick person engages my *sympathy*.

So a part of the point that we should see ourselves as belonging to natural communities could be to urge that widening of our sympathies that others have urged from other directions.[12]

So far I have discussed the way in which environmentalists have taken from ecology scientific support for the notion of community. But the idea of the ecosystem may be seen as significant in the context of the discussion of Paul Taylor's views. Ecosystems are goal-directed systems in the sense employed there, so that if goal-directed systems have a call on our respect, it seems implied that we owe respect to lakes and forests and sand marshes as well as to bacteria and algae.

Summary

- This chapter has explored the support the science of ecology appears to offer to the idea that human beings belong to communities consisting of all the animals and plants, and the habitats of those animals and plants, living in a particular locale. The importance some writers attribute to these communities begins to develop into a move away from individualism, though as the outline of such an alternative begins to appear, some will not find it at all attractive.

Questions

1 Does ecology offer a perspective on the natural world that is helpful in thinking about the place of the human being in nature?
2 On what basis can we say we belong to a 'natural' community?

Further reading

Baird, J. Callicott, *The Conceptual Foundations of the Land Ethic*, Madison, WI, 1987, University of Wisconsin Press.

Leopold, Aldo, *A Sand County Almanac*, 1949, revised edn, Oxford, 1977, OUP.

Merchant, Carolyn, *Radical Ecology*, New York, 1992, Routledge and Chapman & Hall.

—— (ed.), *Ecology: Key Concepts in Critical Theory*, Atlantic Highlands, NJ, 1994, Humanities Press.

Rolston, Holmes, III, *Philosophy Gone Wild*, Buffalo, NY, 1986, Prometheus Books.

Worster, Donald, *Nature's Economy – A History of Ecological Ideas*, Cambridge, 1977, CUP.

9 The importance of being an individual

We have encountered the question of whether or not such and such can be regarded as an individual in a number of contexts. 'Deep Ecology' suggests that we and nature should be regarded as one. Ecology pioneer F.E. Clements said that we should see a developing 'formation' as not a collection of organisms but as a super-organism in its own right. Individualism insists that there is nothing to a society beyond the individuals making it up. The Gaia hypothesis proposes that the earth is one enormous individual, a single organism of which again human beings would have to be thought of along the analogy of component cells.

Plate 9.1 *Is the earth a community of living things? Or just their home? Or an organism itself?*

Source: The earth from space, as appears for example in Colin Blakemore, *The Mind Machine*, London, 1988, BBC Books p. 23. NASA/Science Photo Unit. © NASA

I have presented science as contributing two ideas that many have found useful in trying to think through the environmental situation, the biological idea of an organism, and our common biological provenance, and the idea of an ecosystem. Here I present a third, this time drawn from artificial intelligence.

What then is it to be an individual thing? Clements is saying that a number of individual animals and plants make up a new 'super' organism: what is to be understood as the difference between a group of individuals, and an individual made up of 'components' (which we otherwise think of as individuals themselves)? It's an abstract and difficult question, and in pursuing it one is dogged by the recurring thought: does it actually matter? For the Deep Ecologist however it certainly does seem to matter. They present their insistence that the human being and nature are one not two (or several) as of great importance.[1] And the claim that the earth is one large organic system rather than a large collection of smaller ones was received, at least at first, with consternation. It is partly to keep hold of this question mark over the significance of such an abstract issue that I approach it historically. There *is* a framework of ideas within which the issue of whether something is an individual or not is of acknowledged significance: the framework of medieval Scholasticism, the framework overthrown by Modern science.

Being a thing and possessing a *form*

In fact the question of what was and what was not an individual appeared high up on the agenda when the old medieval way of thinking came under question with the emergence of Modern science. Quite fundamental categories of thought were recast at this time, and some of the recastings continue to provoke disquiet. One concern was this: if a new physics was to make progress, it would have to be based on a correct account of what was *fundamental* in the universe. Should it assume that everything was made up of the tiny inert particles called atoms or corpuscles? If so, all its explanations would be a matter of showing how the phenomena of nature followed from this assumption, together with what further assumptions were made about the character of these 'elementary particles'. Or should it assume that everything was made up not of atoms but of some kind of amorphous 'material'? Or was the correct assumption rather that the universe was made up of fundamental items that were more like *animals* than atoms?[2] In that context, the question of what the difference was between a collection of things and a thing became central. If an x were a collection of things physics would have to analyse it into the individuals that made it up and explain the phenomena associated with it in terms of the characteristics of those members. But if the x were not a collection but a thing 'in its own right' physics would be likely to go wrong if it proceeded on the assumption that it was a 'group' or 'collection'. 'Reducing' the x to the aggregate of its 'parts' would be mistaken.[3]

The opposition to the 'atomism' of early Modern science came partly from the Scholastic philosophy that had been taught in the universities from the eleventh century onward, and which was the common target for the variety of conceptions for a new science that jostled for ascendancy as the late medieval perspective fractured in the seventeenth century.

The notion of a 'form' I have mentioned earlier (see Chapter 1) when I was trying to explain the Scholastic notion of perception. It had been introduced by Aristotle, though it is difficult to get clear exactly what he thought. The Scholastics took it over however, and for them it supports the idea that an individual thing – e.g. a horse – is *matter* (such as flesh and bone) *organised* in a certain way. The idea of 'form' is to a degree like the modern idea of 'organisation'. A horse is made of various sorts of tissue, but these materials are not just in a heap. They are *organised* in a particular way, and that is what makes that particular set of materials into a horse. An individual thing like a horse therefore, according to the Scholastic scheme, is a set of materials, or 'parts', united under a *form*.

So in outline the Scholastic answer to the question of what made the difference between a group of living things and a single 'super-organism', thought of as made up of a large number of component living things, was that the single super-organism would have to have a *form* that in a mere collection would be lacking. The members of a group of living things were thought of as each possessing, individually of course, a *form* – the *form* was what made each member an individual thing. But the group, if it was just a group, was thought of as *not* having a *form* in its own right.

So for the Scholastics the question of what kind of thing a group of organisms was depended entirely on their notion of *form*. What more then can we say about it? I have already said that it is at least a bit like the modern *organisation*. But besides being the organisation it was in some way the *source* of the organisation. In the growth of a plant, for example, or the development of an animal towards maturity, the form was thought of as initiating and guiding the changes involved. In this way you could see the *form* as encompassing the 'nature' of the developing organism. In the sense that it is the 'nature' of an acorn to develop into an oak, the *form* encompasses its nature.

In the sense that it is the nature of the dog to hunt in packs, once again, because the *form* guides the changes in materials that eventuates in a creature with that predilection, the *form* of the dog encompasses its nature.

Here is Roger Woolhouse emphasising the breadth and importance of the Scholastic concept. For the oak tree, he says, its *form* encompasses

> its various parts and their purposes, such as its leaves and bark and their functions; its characteristic activities, such as growth by synthesising water and other nutrients, and its production of fruit; its life cycle from fruit to fruit bearer. It is in being organised and active in this way that the matter which constitutes an oak 'embodies' or is 'informed' by the substantial form 'oak'; it is only by virtue of this that it 'forms' an oak tree at all.[4]

The importance of the role of the *form* in a living thing rules out our understanding it as *simply* the thing's 'organisation'. If the form is somehow responsible for the mature shape, structure and character of the organism, it is playing the role of something that brings about a particular organisation and cannot be equated with the organisation itself. How then are we to conceive of it?

Souls

Thomas Aquinas, the foremost exponent of Scholasticism, thought of the *form* of a human being at any rate as their *soul* and this gives us something we can at least half get hold of today. For people today sometimes appear to think there *is* such a thing as a soul, which they think of as a kind of non-physical entity associated with the body and as somehow carrying the identity of the person. They often make a distinction between the *soul* and the *mind*, but the concept of the mind, as something real but non-physical lodged in the body, appears to serve as a kind of model, or at least a kind of *precedent* for their idea of the *soul*. (Plenty of thinkers in the Modern period have simply equated mind and soul, Descartes amongst them.)

I have said that the soul is conceived of as somehow carrying the identity of the person it 'belongs to', and this aspect of the soul is of course terribly significant. It is the soul, according to Christian thought, and to all the dander that Christian thought leaves in the *common-sense* of the post-Christian West, that transcends mortality, so that having a soul is the most important, most fundamental, most significant fact about a being. If a thing hasn't a soul, it belongs without qualification to this world. Here it comes into being, it has its being here, and here, without passing on to any other world, or realm or order or dimension, its existence ceases. But for a being with a soul this world is a temporary habitation, a gateway to another, a small leg of a journey that will last much (because infinitely) longer. The soul therefore brings with it the conception that life as we know it has a larger context.

This in turn has implications for how we approach morality. Thinking out how we ought to spend our lives, and how we ought to behave in relation to other things, is likely to be much affected *first* by whether or not we think of ourselves as likely to exist for ever, and *second* by whether we think of the things caught up in our actions as short-lived or eternal.

There is another dimension of the concept of *soul* that has a bearing on our conduct. If a being has a soul, there are ways of damaging or harming it to which something without a soul would not be subject. For example, the act of felling and burning a tree may appear to be one thing if the tree is taken to be nothing but a 'collection' of atoms, and quite another if the wood that is consumed in the flames is considered to be the habitation of a soul.

We may not be clear exactly what damage is done in the second case, but there would appear to be a dimension that escapes the 'reductionist' perspective. How are we to understand the idea of damaging a soul? The image that presents itself if we talk of destroying a soul's 'habitation' is of a person *suffering*. We think of a person being driven out of their house, and the mental, perhaps bodily, pain this occasions. In doing so, we are drawing on the link I pointed out earlier between the soul and the mind. We are thinking of the soul as something that is capable of feeling, and in particular of suffering.

In developing the idea that the soul of a human being was his or her *form*, Aquinas was ensuring the concept of *soul* had a wealth of well-articulated meaning. There was much learned discussion exploring the boundaries and certain internal regions of the concept of *form* but its heartlands were regarded as well understood. But all that was lost when the *form*-based framework – the 'hylomorphic' framework, as it is known – was dismembered in the seventeenth century and its place taken, ultimately, by the 'mechanistic philosophy' we are more familiar with in an attenuated form today. It is surprising perhaps that the concept of *soul* survived the revolution. It could no longer be understood as a *form*. There were attempts to identify it with the newfangled *mind* (by Descartes, for example), but somehow orthodox Christian theology (Roman and other sorts) has been able to keep alive a concept of the soul as something distinct from both body and mind. It is a not-body, a not-mind, which carries human identity and that transcends mortality.

In explaining how the term *soul* once had a firm meaning (in virtue of its identification with *form*) I am aware that not a great deal of light is shed on what we may legitimately make of the term today. This is a pity, because some contributors to the environmental debate place a great deal of weight on the notion of a soul. What I want to bring out is that the notion of a soul is not intrinsically vague or ill-defined or shrouded in mystery. It had a reasonably clear sense once, a reasonably clear location in the established conceptual scheme. The sense generated today of a shadowy, mysterious entity of ill-defined features and qualities may simply come from the fact that its conceptual supports have been lost. Those who attempt to use it in the face of this radical isolation are then to be seen as guilty of anachronism, rather than as champions of ideas that deserve deference.

What is to be said then of the 'super-organism' idea? – that what are regarded ordinarily as groups of organisms should sometimes properly be regarded as components of a complex unified whole.

The Scholastics, I explained, had an answer to the question of what the difference was between a single integrated thing – a 'unified' thing, as Clements puts it – and a collection of things. And I have said that corpuscularianism had no answer, arguing that the distinction was not really there – that everything that wasn't an atom was a collection of atoms.

Are we perhaps corpuscularians in principle even today? We think of elementary 'particles' as hardly particles at all, of course – fuzzy not just round the edges but through and through. But nothing else is *less* fuzzy – i.e. we still think there are a small number of different kinds of elementary items out of which everything else is made. This at least would explain the feeling that many of us have that the question of whether a thing is an individual or a part of another individual – or indeed both – is not very clear, and not of the greatest significance.

Individuals and computer programs

Earlier (see Chapter 7) I explained the notion of a goal-directed system and the way of thinking about living things that this offers. Does this notion perhaps also give us ways of distinguishing between organisms on the one hand and groups of organisms on the other? May we not say that a super-organism, were there such, would be something that had a *goal*, or *goals*, over and above any goals that any of its 'components' might have?

We can make this quite concrete. There are in nature both unicells – simple animals like the amoeba, for example – and complex organisms, which we think of as made up of simpler 'cells'. Some organisms change from one 'form' to another from time to time – for example the slime mould, *Dictyostelium discoideum*. Single cells of *D. discoideum* exist for part of the time as independent units, but on receiving a chemical signal (a substance called *acrasin*), the unicells congregate to form what then behaves for a time as a single multicellular entity (see Plate 9.2). (This then, usually, disaggregates, with individual unicells going their separate ways.)

If one just thinks of the *movements* of the different entities here, isn't it appropriate to speak of a new goal coming into play when the composite entity is formed? Before they merge, the unicells move about in different directions but, as components of the new concern, they move in a highly co-ordinated way. Indeed, it becomes somehow awkward to speak of the component cells 'moving' as such: one rather speaks of the moving being done by the new entity.

So may we say this? A super-organism – or multicellular organism as distinguished from a group of unicells – would have a goal or goals apart from any goals that its components might have. But as far as a mere *group* of individual organisms were concerned there would only be the goals of the members.

Today we have our understanding of computing machinery as a model for how we might think of goals and how goals may be related to each other. We may think of an organism as a program, for

Plate 9.2 *Slime mould:* **one** *organism or* **many?**

Source: Half-tone photograph of aggregating amoeba, as appearing on p. 8 in *From Cell to Organism*, intro. Donald Kennedy, San Francisco and London, 1967 edn, attributed to Kennet B. Raper of the University of Wisconsin. © W.H. Freeman and Company

example. If we do that we have no difficulty in thinking of organising a plurality of programs under a super-program. In computer terms the question is simply: Is there a higher goal of which the highest goals of the components are subgoals? In computer terms any such higher goal must be a piece of code. Such a piece of code could be in the components – the whole could be nothing but the components. But there would be 'a unitary thing with parts' if there is a goal to which the goals of the parts are subservient.

This is easily imaginable in an AI (artificial intelligence) context. You can write a program to get a robot to pursue a light source for example, and subsequently place that program within a more comprehensive one that invokes the pursuit of light sources as and when this is deemed appropriate in the light of a more general goal of protecting itself against dangers (some of which may lessen as illumination increases). The robot of your preliminary project has a goal of its own, but when its program becomes a module of some larger enterprise it is the more sophisticated program (and the robot it runs) that has the goals.

This way of thinking about organisms is valuable, it may be argued, because it emphasises that you can have goal-directed behaviour without 'thought'. Clements thought that you can only have goal-direction in a system if it is alive. Many people have reacted to the Gaia hypothesis, which is, in its most interesting version, the proposition that the earth as a whole is a goal-directed system, as though it were being suggested that the earth were not only alive but possessed of 'consciousness'.

The idea that there is a link between goal-direction and consciousness is somehow a very compelling one, but with the development of increasingly sophisticated goal-directed systems that we ourselves have engineered must gradually lose its grip. This should lower the barrier that has been assumed, since Descartes, to separate human beings, who enjoy consciousness, from large tracts of the rest of nature (if not from it all).

The environmental crisis is life and death for many, and maybe for all of us, so that the damage that current environmental policies is doing to the looks and feel and smell and taste of our surroundings may seem of small significance. But this may be a mistake. Who knows what will save the world? Thought concerning the nature of beauty and ugliness has traditionally been targeted on pictures, sculptures and so on, but is now swivelling energetically to consider also the beauty of landscapes and other natural forms (as well as visual values in the 'built environment'). I have asked (another) colleague to offer as a closing chapter an outline of these developments.

Questions

1　What is the difference between living in the natural world and being part of it? Does it matter?
2　Is a forest one thing or many?

Further reading

Bateson, Gregory, *Mind and Nature*, London, 1979, Wildwood House.

Capra, Fritjof, *The Turning Point*, London, 1982, Wildwood House.

Dennett, Daniel, *Darwin's Dangerous Idea*, London, 1995, Allen Lane.

Hofstadter, D.R. and Dennett, D.C., *The Mind's I: Fantasies and Reflections on Self and Soul*, New York, 1981, Basic Books.

Kenny, Anthony, *Aquinas on Mind*, London, 1993, Routledge.

Sherrington, C.S., *Man on His Nature*, Cambridge, 1951, CUP.

10 The aesthetics of the natural environment

While walking in a forest, we might be interested in identifying the trees and birds we see, or we might be interested in the geological or human history of the place. But there is a further dimension to our experience of the natural environment alongside these more knowledge-oriented interests. This dimension is characterised by *aesthetic* interest. Within philosophical aesthetics, aesthetic interest is typically understood as valuing objects or a set of objects in virtue of their aesthetic qualities, such as the sublime darkness of a dense pine forest contrasted with the soft, almost comforting feeling of pine needles underfoot. Environmental aesthetics[1] is concerned with problems and issues that arise in this context of human experience, and the questions it addresses have their roots in ongoing debates within philosophical aesthetics. Some central questions are: What is the foundation of aesthetic appreciation of nature?; Is it science, or something else, such as perception or feeling?; What differences are there between aesthetic appreciation of art and aesthetic appreciation of nature?; Are aesthetic judgements of nature subjective or objective? In this chapter I shall examine these questions first by providing some background to them through a discussion of Immanuel Kant's aesthetic theory. From there I next consider differences between art and nature in aesthetic appreciation. I then turn to a discussion of the current debate within environmental aesthetics between science and non-science-based models of aesthetic appreciation of nature, and how each model handles these questions.

Kant and the foundations of aesthetics of the natural environment

Kant was the first philosopher to write extensively about aesthetic experience of nature. His *Critique of Judgment* (1790),[2] the third of his three 'critiques', examines the nature of aesthetic judgement (and also teleological judgements, which are less relevant to aesthetic theory). Although in this text he is also interested in our aesthetic judgements of art, he begins with examples from nature, and there is good evidence that he viewed nature as the paradigm of an object or set of objects that evoke the aesthetic response. For this reason, as well as his significant impact on philosophical aesthetics, his analysis of both aesthetic experience and judgement is particularly relevant here.

Kant wants to solve the problem of how aesthetic judgements, or what he calls 'judgements of taste', are possible. He begins with an analysis of the nature of this type of judgement, and then proceeds to discover and defend the grounds for such judgements. Through his analysis and argument, an aesthetic theory emerges that shows what is distinctive about aesthetic experience, and how we can make aesthetic judgements that are not merely subjective.

The 'judgement of taste' (which I shall hereafter refer to as 'aesthetic judgement') is expressed in utterances like, 'This poppy is beautiful', or 'That seascape is beautiful'.[3] Kant analyses this type of judgement by setting out its distinguishing qualities. First, aesthetic judgement is contrasted with cognitive judgement because it is not grounded in knowledge of the object, but rather in a feeling of delight or liking in the appreciator.[4] The poppy is judged as beautiful not in virtue of my concept of it as a particular type of flower, nor in virtue of my knowledge that it is the source of opium. I call it beautiful because the poppy evokes an immediate feeling of pleasure, which is a response unmediated by a concept of the object.

The immediacy of the aesthetic response identifies a second feature of aesthetic judgement. If my judgement does not stem from knowledge of the object, then what is the basis of it? I find the poppy beautiful in virtue of an immediate response to its perceptual qualities: the vibrant orange colour, combined with the floppy, tongue-like petals that surround the dark black centre of the flower. But although my enjoyment of these qualities underlies my aesthetic appreciation of it, Kant does not believe that beauty is an objective quality of the flower. Beauty cannot be identified as the orange colour or rounded forms, or its symmetrical design, and these qualities do not alone cause the aesthetic response. That I find the flower beautiful is a direct consequence of an accordance or attunement between the perceptual qualities of the flower and the mental powers that Kant calls the imagination and the understanding. This attunement is characterised as a harmonious 'free play' of imagination and understanding. Freed from conceptualisation, the mental powers engage with the perceptual qualities of the object in a pleasurable activity that is directed at appreciating the object for its own sake.[5] Kant thus characterises beauty as the appreciation of something through an immediate encounter between an appreciator and a particular object. For this reason, we have no predetermined concept of beauty; it is something that arises in a relationship between subject and object.[6]

With these points in place, we are now in a position to more clearly understand why Kant thinks that aesthetic judgement is based in a feeling in the subject, and hence why he says that this type of judgement is subjective. The basis of aesthetic judgement is subjective, rather than objective, for two reasons: (1) aesthetic judgement stems from a feeling in the subject rather than an (objective) concept of the object; (2) as a judgement of beauty, aesthetic judgement cannot be other than subjective, since beauty is not an objective quality, but rather something that arises through an immediate encounter between appreciator and object. These points reveal the particular way in which Kant wants to identify aesthetic judgement as subjective, and they also introduce a vital point. It would be a serious mistake to associate Kant's claim that aesthetic judgement is subjective with the view that 'everyone has his own taste' or that 'beauty is in the eye of the beholder'. Kant adamantly denies that this is the case with objects that we find beautiful, and one of the main arguments of the *Critique of Judgment* is to show that aesthetic judgement has a subjective basis in feeling, yet it nonetheless claims universal validity. When I find the poppy beautiful, I am not making a singular judgement that applies only to myself, but one that assumes universal assent.[7]

Although Kant's defence of this claim is too complex to discuss in detail here, one aspect of the argument is particularly relevant because it highlights another important feature of aesthetic judgement, 'disinterestedness'. In this context, disinterestedness is not indifference, or lack of interest. Kant uses the concept to characterise aesthetic interest as distinct from interest in an object as a means to sensory gratification, and an interest in using it as a means to some utilitarian end.[8] The disinterestedness that characterises aesthetic judgement issues in a type of appreciation that appreciates the aesthetic qualities of an object apart from any end. My appreciation of a seascape is aesthetically disinterested, according to Kant, when it rests on valuing its aesthetic qualities, e.g. the graceful underwater movement of seals against a striking blue backdrop, rather than valuing it as a place to refresh myself after sunbathing (for sensory gratification), or as a mineral resource (where it serves a utilitarian end).

Disinterestedness operates in aesthetic judgement to distinguish this type of appreciation from the arbitrariness and subjectivity of personal desires, as well as other more general aims that are irrelevant to the aesthetic approach. As a condition of aesthetic judgement it secures impartiality, by freeing ourselves from personal desires or preoccupations with utility in relation to the object we are in a better position to judge the object on its terms. And if this condition holds, it becomes clear how disinterestedness provides some support for Kant's claim that aesthetic judgements are universally valid. When I approach the poppy or the seascape disinterestedly, and assume that others do so too, then I have a priori grounds for expecting the agreement of others when I judge something to be beautiful. (Kant offers further support for his argument by claiming that we share similar perceptual and cognitive apparatus, and thus we all share the basic means to appreciate beauty.[9])

Kant's analysis of aesthetic judgement provides a useful starting point for a critical understanding of what is distinctive about the aesthetic response, and what is involved in aesthetic appreciation of nature. The aesthetic response is distinct from other ways in which we relate to the world because it is based on an immediate, disinterested feeling of delight in response to the perceptual qualities of an object. For Kant then, knowledge is not the basis of aesthetic appreciation, but rather the feeling of pleasure that comes through perception of the object and the harmonious free play of the mental powers. The distinctiveness of aesthetic appreciation is further developed by Kant in his belief that aesthetic judgement is neither subjective nor objective. Instead it lies somewhere in between, having a subjective basis in feeling, yet at the same time asserting the agreement of others. Aesthetic judgements have subjective universal validity, which distinguishes them from other types of judgements we make, including cognitive and moral judgements.

As noted above, Kant takes nature as the paradigm of the aesthetic response, and he even argues, if not entirely consistently, that only nature is the appropriate object of the 'pure' judgement of taste. But he is also interested in other categories of the aesthetic response, including our response to the sublime (which is almost exclusively a response to natural objects), and our response to art. I shall not discuss these other aspects of his aesthetic theory here, although his theory of the sublime in particular is of special interest to environmental philosophy because it explores another type of aesthetic relationship with the natural world. I would, however, like to draw attention to one point that is relevant to understanding differences between the aesthetic appreciation of art and nature. Kant recognises the essential distinction between art and nature, which is that the former is the product of human creation. Moreover, as artefacts, artworks are fashioned according to some concept or idea of what the object is meant to be, so in this respect art is distinguished from nature as the product of human intention.[10] The upshot of this for the aesthetic response is that, unlike nature, aesthetic appreciation of art involves, at least in the background of our response, some idea or concept of what the object is intended to be.

Kant's influence in philosophical aesthetics generally has been great, but his aesthetic theory is particularly relevant to our concerns here because it provides the first model of the aesthetic appreciation of nature. The enormous influence of this particular model can be seen in contemporary debates on the aesthetics of nature. Not only is his distinction between art and nature upheld, but even more interesting is how Kant's views have influenced *both* sides of the current debate, despite the great differences that exist between the two most recent competing models of aesthetic appreciation of nature. Before moving on to a discussion of these models, I shall expand on some of the key differences between art and nature that have emerged in more recent work in the aesthetics of nature.

Art and nature

Constructing a useful model of aesthetic appreciation of nature first requires understanding how our approach to nature compares and contrasts with our approach to art. Some have argued that we should view nature through the framework of our experience of art, while others have resisted this move, arguing that it reduces nature to the artistic and ignores essential differences between art and nature. While there are many similarities between our aesthetic experiences of art and nature, I shall concentrate on the differences, since they highlight the importance of treating nature as requiring its own distinctive model of aesthetic appreciation.[11]

One clarification is necessary from the outset. When using the term 'natural object', we should note that the expression refers to any object that is not a product of human creation. However, many things we call natural will originate in both natural and human causes, as is the case with an artificial lake or an agricultural land-scape. For our purposes here, I shall assume that an essential, if not clear-cut, distinction exists between artefacts and natural objects, even when the origin is an integrated one.[12]

The character of our aesthetic experiences of nature versus artefacts may be distinguished by two main features: (1) *the nature* of the object ; and (2) the *context of* our experience of the object. First, the nature of the object necessarily includes the origin of the object. This points to the most essential difference between natural and art objects. An artwork is first and foremost an intentional product. As such it is an object that embodies the ideas of its maker. It follows from this that our response to artworks is shaped and in some sense controlled by the properties of the particular object – the colours, lines, forms and images on a canvas, or the notes and melodies of a tune. Our experience is thus *directed* by cues in the artwork through whatever particular medium we encounter. In this way, the experience is overlaid by an awareness that the object is intentional so that we seek meaning in the object through aesthetic engagement with it. The meanings we seek are there to be found, and we attribute such mean-ings to the artist who has 'purpose-built' the object for our interpretation and enjoyment.

As Kant observed, natural objects are not the products of human creation, and so for this reason we are not given a body of forms as we are with art in which to find meaning through interpretation. We do not view nature as holding certain meanings that stem from the intentions of a creator;[13] we make no attempt to 'read' natural objects in this way. Given this fact about the origin of natural objects, our response is not directed by an artist, but by our own experience and features of the object.

The lack of a creator makes the object interpretively indeterminate, which has both positive and negative implications for aesthetic appreciation. The positive value of the lack of a maker means freedom from the guidance of an artist in terms of his or her intention as it is embodied in the artwork. In a Kantian sense, we are truly freed from all concerns about the kind of thing the object is intended to be, in terms of what it is intended to express or how it functions as an object in the world. There is more scope for various meanings to be brought to nature, and to search for meanings in the object that are tailored by ourselves through what we sense, what we imagine, what we feel emotionally about nature, or by the various narratives that inform our aesthetic encounters (e.g. folklore, natural and human history, myths, and religion).

A second important difference concerning the nature of the object of aesthetic appreciation relates to the fact that artworks are physically bounded objects while natural objects are not. For example, paintings are bounded by frames, and the ontology of artworks is determined by both physical boundaries and conventions. Some natural objects stand out from their environment as single, individual objects for contemplation while others blend with it in ways that make the object of our experience a group of objects, or one vast object that reaches beyond the grasp of our senses. Such objects come without frames, and many are impossible for us to frame through mere perception. But in these examples I have described situations in which we are still able to set ourselves apart from natural objects, viewing them from afar as we do when we stand back from artworks. What is more interesting is not just that natural objects are unbounded, such that if they have any frame it's the frame we place around them through perceptual attention, but rather that we are most often *not set apart* from natural objects. This sort of aesthetic encounter is less common with artworks since we do not typically encounter them as *environments*.[14] Nature offers both individual objects, plants or clouds, but also total environments, the desert or forest, as the focus of aesthetic attention. We find ourselves amidst the object: standing in the middle of a desert, we can turn around slowly to contemplate its colours and the shapes of the sandstorms moving through it. With nature total immersion in the aesthetic object is possible.

That we may become so immersed points to another feature of natural objects: *changeability*. Natural objects are always changing due to growth, decay, erosion, climate changes or the inherent movement of certain objects like the earth or the sea. By our own volition it is possible to take up a variety of perspectives in light of such changes, but more often these changes force new aesthetic perspectives on us. Viewing the Scottish Highlands in rapidly changing weather comes to mind here.

These non-static aspects of natural objects are closely connected to the second feature that reveals the important differences between the natural and art. The *context of* the percipient's experience changes due to the nature of the aesthetic object. Because art objects are physically bounded, our various aesthetic perspectives and engagement with them are determined by those boundaries. Hence the scope of our aesthetic experience is limited. Natural objects present the possibility of a multi-sensuous experience due to the fact that we often engage with natural objects through immersion. Standing on a rock on the edge of a turbulent sea offers an active aesthetic encounter that emerges through sheer bombardment of the senses – the mist thrown up by the waves, the wet, fresh smell of the sea, the sound of the crashing waves, and the taste of salt in one's mouth. This contrasts sharply with the static, yet perhaps equally exciting, experience of watching a film in a dark cinema.

Furthermore, we can move through the aesthetic objects we experience. Standing still to contemplate a wildflower is one way to appreciate the environment. But we are also able to walk in it, climb it, fly through it, sink into it, and swim through it.[15] The potential for a multi-sensuous experience also suggests the possibility of a more dynamic response in our experience of natural objects. That natural objects are unbounded, non-artefacts, changeable, and offer the experience of multi-sensuousness and total immersion makes them 'aesthetically inexhaustible'.[16] That is, the indeterminacy of such objects widens the scope of our aesthetic experience in equally indeterminate ways. The complexity of such objects provides the possibility of rich and rewarding aesthetic experience, but the aesthetic response to nature is not given. We will not appreciate nature unless we take up the aesthetic challenges it presents to us, as argued by Ronald Hepburn:

> we can contrast the stereotyped experiences of the aesthetically apathetic and unadventurous person with the rich and subtly diversified experiences of the aesthetically courageous person. His courage consists in his refusal to heed only those features of a natural object or scene that most readily come together in a familiar pattern or which yield a comfortingly generalised emotional quality.[17]

This plea for exploring the possibilities of nature aesthetically, to try out fresh perspectives rather than viewing it through the car window or a camera lens, points to a practical implication of this realm of our experience. Our aesthetic encounters with nature can often deepen our attachment to nature in ways that are not instrumental, and in ways that engender greater respect for the natural environment.

The differences between art and nature appreciation cannot be ignored in understanding what is distinctive about the aesthetic response to nature, but at the same time they create some difficulties. For example, although it is the very indeterminacy of natural objects and the dynamism of our aesthetic response to them that may lead to a rich and exciting aesthetic encounter, these aspects make it more difficult to pin down clear guidelines for appreciation.

When we appreciate an artwork, such as a painting, the perceptual qualities of the work guide our visual and imaginative exploration of the canvas. But this experience may also be guided by what we know about the history of the painting and its artist, as well as by the individual experience we bring, including particular associations and any emotions that are evoked by the artwork. I might appreciate Van Gogh's *Sunflowers* through enjoyment of its colour and form, but it is more typical that my interpretation and evaluation of the painting will be determined according to what I see supplemented by cultural and historical knowledge.

When we turn to nature, however, aesthetic appreciation lacks the guidance of an artistic context. Natural objects lack a human maker, an artist, and also an artistic context in respect of the type of artwork, e.g. painting or novel, and in respect of style, e.g. cubist or surrealist. We are familiar with some of the ways in which the aesthetic response to art is guided by features of both the work and the individual appreciation, but what guides our aesthetic appreciation of nature? Our aesthetic response to the seascape is guided by what I see, colours and shapes, or what I grasp with the other senses, as well as background knowledge and individual experience we bring to the experience, such as the narratives of folklore or memory, but it lacks an artistic context provided by an artist or a body of artworks. This context is significant to the appreciation of art because it identifies some means by which to fix appropriate appreciation. For example, Kendall Walton has argued that appropriate aesthetic appreciation of art must be guided by knowledge that enables us to perceive it in the correct category, so that we appropriately appreciate *Sunflowers* if we perceive it in the category of post-impressionism rather than cubism.[18] Once we perceive it in the correct category, we are better able to determine whether it is a good work of art or not. (His point is aimed against the narrowness of formalism, which argues that the perception of form, design and colour is sufficient for the interpretation and evaluation of art.) Although Walton's position is not unproblematic, it points to how the artistic context provides at least some foundation for the appreciation of art.

The comparison between art and nature highlights the problem that arises when artistic context is absent from aesthetic appreciation: What replaces artistic context in the appreciation of nature? What frames our aesthetic interpretation and evaluation of poppies and seascapes? For Kant this was not a problem, since he begins with nature as the paradigm of aesthetic appreciation. After Kant and until more recently, these questions were answered in terms of a landscape or scenery model of appreciation, which holds that appropriate aesthetic appreciation of nature is achieved by viewing nature as if it were a landscape painting. The appropriate approach is defined according to the standpoint we take to landscape paintings, where we stand back and behold the design, forms and colours of the picture. Moving from the art gallery to the natural landscape, we stand in one place and enjoy what we see as a scene, a canvas laid before us, bounded not by a wooden frame but by the horizon, and the limits of the visual field.[19] The landscape model is rather outdated now, mainly due to the way in which it privileges art to nature, as if the only way we could appreciate nature is through the lens of art. This problem for the model rests on two mistaken assumptions.[20] It assumes that nature is appropriately appreciated in

Plate 10.1 *From the art gallery to the natural landscape*

Source: *Picture* by Magritte

the category of pictures, i.e. two-dimensional, fixed images on canvas or photographic film. But to view nature as a picture narrows appreciation to aesthetic features such as design and colour, and although they can sometimes be relevant, they are unnecessarily restrictive on aesthetic experience.[21] The differences between art and nature discussed above show that nature offers the opportunity of more dynamic appreciation in terms of its changeability and the possibility of aesthetic immersion in the natural environment. Moreover, a second assumption implicit in the landscape model is the view that nature has aesthetic value only when it is framed by art. This is objectionable on both ethical and aesthetic grounds, for it does not recognise that nature is worthy of appreciation in its own right; to view nature as a work of art fails to respect nature for what it is.

The current debate: science-based and non-science-based models of the aesthetic appreciation of nature

More recent debate on the question of appropriate aesthetic appreciation of nature moves us forward from criticisms of the landscape model. The various theories on each side of the debate can be categorised into what I call the 'science-based model' and the 'non-science-based model'. The set of theories I designate as science-based fall into this category because they argue that we should turn to scientific knowledge – ecology, geology, etc. – to guide aesthetic appreciation of nature.[22] The non-science-based theories argue that scientific knowledge is not essential for such appreciation, and they put forward an alternative framework or basis for appreciation, consisting in immediate perceptual experience, imagination and non-scientific narratives, among other features of experience.[23] The distinction between the two models does not rest on an opposition between objectivity and subjectivity, for although the science-based model strives towards objectivity in aesthetic judgements of nature, the non-science-based model is not opposed to this aim either, although it is more sympathetic to the role of subjectivity. The distinction is also not sharply drawn along the lines of cognitive content. Knowledge of one kind or another (scientific, cultural, historical, religious) plays a role in each of the two models, and in this respect there is certainly some overlap between them. The non-science-based model argues that scientific knowledge might play some role but that it is not necessary for aesthetic appreciation, while some science-based approaches argue that non-scientific knowledge has some role to play in aesthetic appreciation.

I begin a closer examination of the two models by focusing on the most established of the science-based approaches, Allen Carlson's 'natural environmental model'. Carlson draws on Walton's argument to contend that knowledge of the natural sciences and their 'common-sense predecessors and analogues' replaces artistic context in our appreciation of nature. Based on this strategy, he argues:

> The analogous account holds that there are different ways to perceive natural objects and landscapes. This is to claim that they, like works of art, can be perceived in different categories, but not, of course, in different categories of art, but rather in different 'categories of nature'. Analogous to the way *The Starry Night* might be perceived either as a post-impressionist or as an expressionist painting, a whale might be perceived either as a fish or as a mammal…Further, for natural objects or landscapes some categories are correct and others not.[24]

This position identifies natural science and its 'common-sense predecessors and analogues' as a replacement for the knowledge of artistic traditions and styles that guide our appreciation of art. What follows from this is that appropriate and significant aesthetic appreciation of nature is impossible without 'something like the experience and knowledge of the naturalist'.[25]

Carlson does not assume that all aesthetic responses to nature will be grounded in what is a rather strong epistemological requirement on the appreciation. He acknowledges that the aesthetic response comes through the mere perception of our environment, the careful use of which will lead to the discovery of aesthetic qualities in nature. The perceptual element is explained through 'order appreciation' in which the appreciator's perception selects aspects of the natural object, and 'focuses on the order imposed on these objects by the various forces, random or otherwise, which produce them'.[26] The exercise of selecting and focusing is guided by a non-aesthetic and non-artistic story, provided by science, and it is this story that enables us to perceive that order. Thus, perception in the aesthetic response to nature is dependent on scientific knowledge. Moreover, science not only serves to deepen aesthetic appreciation, but without it we are unlikely to make aesthetic judgements that are *true*. We may overlook or misapprehend some aesthetic quality in nature unless aesthetic appreciation is supported by scientific categories, as Carlson's own example illustrates:

> The rorqual whale is a graceful and majestic mammal. However, were it perceived as a fish, it would appear more lumbering, somewhat oafish, perhaps even a bit clumsy (maybe somewhat like a basking shark).[27]

The use of science to ground aesthetic appreciation of nature thus provides one solution of how aesthetic judgements of nature can be objective. By combining perceptual qualities with the objectivist epistemology of science, Carlson is able to pin down something close to general criteria for the aesthetic evaluation of nature. What we know about nature informs our perception of it by providing categories through which to correctly perceive it. I would be wrong in my judgement that the whale is clumsy if I perceive it in the incorrect category, just as I would be mistaken if I judged one of Picasso's cubist works to be a poor attempt at impressionism.

Carlson is critical of aesthetic subjectivism for good reasons, and this is surely a strength of his account. The ethical implications of aesthetic subjectivism are potentially quite dangerous to the natural environment. If beauty is in the eye of the beholder, then it could be argued that there is no way to arbitrate between personal opinions. One person's view will be as sound as the next person's, so that, for example, a nature conservationist who wishes to prevent a road being built through a forest will have no defensible aesthetic grounds for arguing their case. Carlson's broad aim is to show that we can make aesthetic judgements of nature that have some claim to truth, and if aesthetic value is to play any role in environmental decision-making, then it cannot be reduced to the arbitrariness of extreme subjectivity.

Carlson's use of scientific knowledge makes sense as a foundation for an aesthetic response to nature. If one agrees with Walton's argument, it is reasonable to appeal to *natural* history instead of art history to determine appropriate appreciative categories for nature. As artefacts, paintings can be contextualised according to their history, and for natural objects, rocks or forests, why not turn to their history, such as geology and ecology? But his emphasis on natural science leads to a number of criticisms of his view. These criticisms are rooted in the more general objection that science as the foundation of aesthetic appreciation excludes some very common ways in which we respond to nature aesthetically. For example, we respond to nature through mere sensory engagement that is free from the epistemological constraints of science. This mode of aesthetic attention can serve as a solid foundation or guide for appreciation, and it can also achieve the depth appreciation that Carlson seeks through knowledge. Alongside this alternative lies another: that the aesthetic response to nature involves being *moved* by nature. Feelings and emotions are stirred by the call of the curlew or the sound of wind through a pine forest.

One criticism that emerges from the general objection is the claim that the scientific model supports a disengaged approach rather than embracing the distinctive possibilities offered by the natural environment. Carlson recognises that nature is an environment, and not merely an object or set of objects, but his model retains the distanced, contemplative and rationalistic aspects of scientific analysis. This criticism has been put forward by Arnold Berleant who argues that the traditional disinterested standpoint of aesthetic appreciation is mistaken, even in the context of art.[28] Rather than standing back from nature, he holds that perceptual immersion is more suited to the aesthetic approach, and more suited to appreciating nature as *environment*.

Related to this criticism is the view that Carlson over-emphasises the cognitive content of the aesthetic response to nature. To some it seems counter-intuitive to claim that knowledge is essential for appreciating nature aesthetically, and, in particular, knowledge of science. Scientific knowledge is a good starting point for appreciation that is characterised by curiosity and wonder, but it is less clear that it should be essential for *aesthetic* appreciation. For example, we can appreciate the smell of pine needles or their muffled softness underfoot without knowing what tree species they belong to. Discovering their aesthetic qualities relies only upon perception and feeling. Another reason given in support of this criticism is the belief that knowledge can impede attention to the appreciation of perceptual qualities, and thus divert attention away from the aesthetic and towards an intellectual mode of attention. These points have their roots in Kant to some extent, since they implicitly pin down the traditional components of aesthetic experience. However, it is worth noting that this criticism does not argue for a formalist approach that seeks to make all knowledge inappropriate. Rather, it is critical of making any type of knowledge, especially scientific knowledge, a *condition* of appropriate aesthetic appreciation of nature.

In an attempt to respond to this type of criticism, Carlson has tried to set out more clearly what is involved in his epistemological condition. But this brings two related problems to the surface. First, he attempts to generalise the scientific categories in a way that reduces them to everyday knowledge, but this weakens their usefulness for the aims of his model. Second, if we abandon generalising them and return to a stricter use of his categories, the model becomes both overly restrictive on the aesthetic response, and it leaves one wondering what is left of the *aesthetic* in his approach.

Other science-based approaches are more open to the range of aesthetic responses we have to nature, and so they have the advantage of being more inclusive. For example, Holmes Rolston and Marcia Muelder Eaton have argued that both literature and personal narratives are appropriate frames for appreciating nature aesthetically. But they also insist on science as the basis of this appreciation, and they are wary of a significant subjective component in our response.[29]

Many of the problems of the science-based approach were in some sense anticipated in an important article by Hepburn first published in 1969, 'Contemporary aesthetics and the neglect of natural beauty'.[30] Here he argues that philosophical aesthetics had become a philosophy of art, and he works towards rectifying that through an aesthetics of the natural environment. In the article he provides not only the first instance of a non-science-based approach, but also a set of original ideas that, as it turns out, suggest ways to address the main problem of the science-based approach, its narrowness and incapacity to cover the diversity of aesthetic responses to nature.

Hepburn recognises the myriad responses that are possible in this context, and he argues that they are determined not by scientific knowledge, but by perception, emotion and imagination. Perception enables us to focus on particular details, or to choose a wider sensory field for aesthetic attention, while emotional qualities are grasped through both perceptual and imaginative exploration. The suggestive sensuousness of a red poppy is only grasped when it is experienced up close, and our response may be different altogether if we viewed a field of poppies from the distance.

Through his analysis of perception, imagination and emotion in our aesthetic response to nature, Hepburn also provides an answer to the role of knowledge in our aesthetic response to nature. Scientific knowledge has no fundamental role in aesthetic experience of nature (although it may certainly be part of the background knowledge we bring to the experience). Nonetheless there is an epistemological component in what he calls an experience of 'realising', which 'involves making, or becoming, vivid to perception or to the imagination'.[31] This is best understood through his own example:

> If suddenly I realise the height of a cumulo-nimbus cloud I am not simply *taking note* of the height, but imagining myself climbing into the cloud in an aeroplane or falling through it, or I am superimposing upon it an image of a mountain of known vastness.[32]

As a feature in the aesthetic response, imagination enables us to become aware of various aspects of a natural object. The approach is not objective in the sense that we come armed with facts or information that we use as categories through which to perceive nature. Instead, our understanding of the cloud arises from gazing at it and contemplating its forms through imagination. Hepburn, like Carlson, is aware of the difficulties involved in extreme subjectivity, and the possibility of manipulating nature through self-indulgent fantasy or trivialising it through sentimentality are not part of Hepburn's view. The pitfalls of subjectivity are best avoided, he argues, by keeping our imaginings relevant according to the perceptual qualities of the object and the individual qualities that make that object distinctive. More recently, he has argued that we should avoid an aesthetic approach to nature that 'distorts, ignores, suppresses truth about its objects, feels and thinks about them in ways that falsify how nature really is'.[33] Determining the boundary between true and false approaches is not determined by science, but on a case by case basis, taking into consideration the variability of nature and the individual experience of the appreciator.

A number of other non-science-based theories have emerged since Hepburn's article. There are a variety of perspectives, but all of them downplay an objectivist epistemology as the foundation of the aesthetic response. However, with the exception of Berleant's theory, they do not want to replace Carlson's model but they are critical of it, and they urge adopting a non-science-based model *alongside* a science-based one.

Although I cannot discuss here every theory in the non-science-based camp, it is possible to identify some common themes, some of which are already familiar from my discussion of Hepburn. Many of the philosophers on this side of the debate favour *a phenomenological* aesthetic approach.[34] The phenomenological perspective in this context functions both as a critical and descriptive framework. It is critical of the subject–object and mind–body dichotomy implicit in Kant's model, the landscape model and the natural environmental model. One of the fundamental drawbacks of these models is the assumption that nature is appreciated as an object from the perspective of a disembodied, contemplative appreciator. In one important respect, though, the phenomenological approach harks back to Kant, in that it emphasises the immediacy of perception and feeling in the aesthetic response. But disinterestedness, which in part facilitates the unmediated nature of the aesthetic response for Kant, has no place in this approach. Instead, this approach assumes fullness of participation. Berleant's view brings these ideas together nicely:

> Perceiving environment from within, as it were, looking not *at* it, but *in* it, nature becomes something quite different; it is transformed into a realm in which we live as participants, not observers…The aesthetic mark of all such times is not disinterested contemplation but total

engagement, a sensory immersion in the natural world that reaches the still uncommon experience of unity.[35]

Berleant's aesthetics of engagement values activity rather than passivity, involvement rather than distancing.

Another common theme of the non-science-based approach issues from the idea of engagement. In the move away from a traditional or modernist aesthetic, engagement is accompanied by attention to the context or situation of the appreciator. In this way the experience of the individual appreciator – emotions, values, beliefs and memories – become as important as the object of aesthetic attention. Thus, the aesthetic response, including any judgements we make, is determined as much by the situation of the appreciator as by the natural environment. Consider, for example, the view expressed by a resident living near a forest in Finland through which a new road may soon be built:

> You should never plan a road if you haven't visited the place many times. It is not enough to go there once…You should go in different moods. You should go when you're drunk, and try the feeling of how it is to sing in the forest. You should go the following day when you have a hangover. You should go when your heart is broken…Then perhaps you know if you can build that road or not.[36]

This shows in a direct way what the non-science-based approach hopes for in a practical sense: individual aesthetic experience coming together with an informed understanding of the natural environment, where that understanding is achieved as much through aesthetic sensitivity and involvement as through knowledge. Knowledge, in this case provided by ecologists and developers involved in the decision-making process, lies not at the foundation of the aesthetic response but has a role in supplementing or existing alongside that experience.

A final common theme, also connected to engagement, is the view that feeling and emotion can play a meaningful role in our aesthetic response to nature. Kant's theory emphasises a feeling of delight or enjoyment, but this is something different from the ways in which we might be moved emotionally by nature, or might find emotional qualities in our natural environment. Hepburn, above, refers to the latter type of emotional response, but I have not yet discussed the former. One view, put forward by Noël Carroll (which Carlson refers to as the 'arousal model'), argues that our being moved by nature belongs to the class of being emotionally moved. This point is important to this model because it shows how aesthetic appreciation of nature can include the emotional response without the problems of subjectivity. Carroll argues that emotional responses are deemed appropriate or inappropriate according to their objects and the aspects of the emoter that underlie them, including his or her beliefs and thoughts. Whether or not it is appropriate for me to feel melancholy when standing in the middle of a desolate moor depends upon the nature of this particular moor, but, also, the beliefs and thoughts that underlie the response. Furthermore, we must ask if the thoughts and beliefs which underlie that response might be reasonably shared by others.[37]

These various aspects of the aesthetic response – perception, imagination, emotion – are given both stronger and weaker emphasis in the various theories that fall into the non-science-based model. A notable strength of this model is the way in which it opens out the dimensions of aesthetic experience by including rather than excluding a range of modes of aesthetic attention. This makes the model more flexible and free than the science-based model, so that it is more able to cope with the complex demands of aesthetic appreciation of the natural environment. Furthermore it makes room for and in fact encourages some subjective elements of the aesthetic response, including the situation and context of the appreciator. This last point is of particular importance in the practical context, where attention to detail and localised knowledge is imperative for good judgement. Attention to the aesthetic experience of local residents, coupled with the perspectives of the ecologist and the developer, are all voices that have a role in the deliberative process. The thoughts of the local resident quoted above stem from his individual experience in that forest, thoughts which are both included and valued in the non-science-based model of aesthetic appreciation.

But alongside these strengths lie a number of weaknesses that also affect use of the model at a practical level. Carlson has criticised non-science-based approaches for their lack of cognitive content, pointing to, for example, the sensory immersion that is the basis of Berleant's engagement model.[38] Carlson argues that mere sensory immersion is tantamount to emptiness, that is, to the 'blank cow-like stare' of the disinterestedness that Berleant wants to avoid with his idea of engagement. This argument continues with the claim that without the cognitive content supplied by science, perception, emotion and imagination in varying degrees lead to subjectivity of the 'beauty is in the eye of the beholder' sort.[39] We might project on to nature a set of emotions that it cannot reasonably have from an objective point of view. Or there is the danger that we manipulate or even sentimentalise nature through imaginative fantasy.

Hepburn, Berleant and Carroll have all defended their positions against these charges. And we would be missing the subtlety of their accounts if we assumed that they favour an instrumental approach to nature that makes the pleasure of the appreciator the basis and aim of aesthetic experience. However, the subjective element of the non-science-based model does make it difficult to fix any general criteria in practice. For example, the range of individual experience that would have to be taken into account from a variety of individuals presents the practical problem of incorporating this diversity into the deliberative process.

Aesthetic value and environmental policy

What is the role of aesthetic value in deciding whether or not to construct a new road through a forest, or to use a mountain for a superquarry? What type of model of aesthetic appreciation of nature will best define this role? The most desirable model will solve the problem of how to guide appreciation in the absence of artistic context and provide a way to make aesthetic judgements that are not merely subjective. Also important is the need to carve out a distinctive place for aesthetic value in the deliberative process. The aesthetic response defines one of our most basic relationships with the natural environment, and although it may not always take precedence, it must at least have a place in decision-making. Thus the best model will also make it possible to distinguish aesthetic value from other values such as ecological value or recreational value.

The science-based model frames aesthetic appreciation through scientific knowledge, and in this it achieves some objectivity. This degree of objectivity is also coupled with an impartiality that values nature non-instrumentally through appreciation of aesthetic qualities for their own sake. However, alongside these advantages there are also some disadvantages. Because of this model's emphasis on the role of

Plate 10.2 *Slicing streets into the wilderness*

Source: Colour photograph appearing in *National Geographic* 174(6), December 1988, p. 776. © *National Geographic*

knowledge, scientific and aesthetic value might become indistinguishable in the deliberative process. Furthermore, the narrowness of this model may exclude the important experience of individuals who have aesthetic ties to it that cannot be reduced to objective criteria. The non-science-based model more clearly demarcates aesthetic value by embracing the features of experience we more typically associate with the aesthetic, but with this comes the drawback that aesthetic value is more difficult to measure.

Given the strengths and weaknesses on each side of the debate, what is the best way forward? One way forward is to combine the most desirable features of each model. Total objectivity in aesthetic judgements is an impossible goal, yet we need to find ways to reach agreement in assessing the aesthetic value of some aspect of the natural environment. In the decision-making process, it is desirable to be as inclusive as possible of the range and diversity of aesthetic responses, including both positive and negative ones. This goal is not unrealistic, for much work can be done at the level of discussion between people who are come equipped with a good understanding of the case at hand. The basis of that understanding will range from the expertise of scientists or historians, to aesthetic understanding, a way of knowing the environment through aesthetic sensibility. Settling disputes in this discussion is made easier by excluding those responses that are clearly self-seeking, and here a renewed notion of disinterestedness might be useful. Disinterestedness has the advantage of excluding self-interest and a 'pleasure-seeking aesthetic', but including other aspects of the appreciator, such as the beliefs, values and life experience that shape their aesthetic encounters with nature. By combining the non-instrumentality entailed by disinterestedness with the situated and first-hand experience of the appreciator, we move towards a more viable concept of aesthetic value for environmental policy-making.

Questions

1 Is it possible for a picture of a beautiful landscape to be beautiful in the same sense?
2 How might the importance of preserving beautiful things be properly represented in environmental policy-making?

Further reading

Berleant, A., *Aesthetics of the Environment*, Philadelphia, 1992, Temple University Press.

Eaton, M.M., *Aesthetics and the Good Life*, Rutherford, NJ, 1989, Farleigh Dickinson University Press.

Hepburn, Ronald, *Wonder and other Essays*, Edinburgh, 1984, Edinburgh University Press.

Kant, Immanuel, *Critique of Judgment*, 1790, trans. W. Pluhar, Indianapolis, 1987, Hackett.

Kemal, S. and Gaskell, I. (eds), *Landscape, Natural Beauty and the Arts*, Cambridge, 1993, Cambridge University Press.

Sepanmaa, Y., *The Beauty of Environment: A General Model for Environmental Aesthetics*, Denton, 1992, Environmental Ethics Books.

Notes

2 Objective nature

1 See Richard Rorty, *Philosophy and the Mirror of Nature*, UK edn, Oxford, 1980, Blackwell.
2 For clear and authoritative exposition, see Anthony Kenny, *Aquinas on Mind*, London, 1993, Routledge.
3 Keith Thomas, *Man and the Natural World*, London, 1983, Allen Lane.
4 John Parkinson in 1640 divided plants into the following categories: 'sweet smelling', 'purging', 'venomous', 'sleepy', 'hurtful' and 'strange and outlandish'. (See Thomas, pp. 52ff.)
5 For these examples, see Thomas, pp. 54ff.
6 Thomas, p. 51.
7 Michel Foucault, *The Order of Things*, UK edn, London, 1970, Tavistock, p. 129; emphasis added.
8 *The Philosophical Works of Francis Bacon*, J.M. Robertson (ed.), London, 1905, Routledge, pp. 409–12.
9 Buffon, *From Natural History to the History of Nature*, J. Lyon and P.R. Sloan (eds), Notre Dame, 1981, Indiana, University of Notre Dame Press, p. 110.
10 Foucault, op. cit., p. 40.
11 Foucault, op. cit., p. 35.
12 Carolyn Merchant, *The Death of Nature*, London, 1980, Wildwood House.

3 We are all one life

1 Coleridge, quoted in M.H. Abrams, *The Mirror and the Lamp*, Oxford, 1953, OUP, p. 65.
2 Immanuel Kant (1724–1804), *Kant's Political Writings*, trans. by Hans Reiss, Cambridge, 1970, CUP, p. 54.
3 Michel Foucault, *The Order of Things*, UK edn, London, 1970, Tavistock, Chapter 5.
4 Gottfried Leibniz (1646–1716), letter to Foucher in Leibniz, *Philosophical Writings*, trans. by Morris, London, 1934, Everyman, p. 48.

5 J.G. von Herder (1744–1803), quoted in Roy Pascal, *The German Sturm und Drang*, Manchester, 1953, Manchester University Press, p. 135. 'The great achievement of English Romanticism was its grasp of the principle of creative autonomy' – Northrop Frye, 'Blake after two centuries' in M.H. Abrams (ed.), *English Romantic Poets*, New York, 1960, Galaxy, p. 65.

6 There is a distinction between the German *Sturm und Drang* movement of the last quarter of the eighteenth century and the Romantic movement, which began later. It is nevertheless arguable that the earlier movement had much of the character of a herald to the later.

7 Lenz, trans. in Roy Pascal, *The German Sturm und Drang*, pp. 48, 49.

8 S.T. Coleridge, *Biographia Literaria...with the Aesthetical Essays*, J. Shawcross (ed.), 2 vols, London and New York, 1907, OUP, corrected edn 1954, Chapter XIII, Vol. I, p. 202; and Coleridge, *Collected Letters*, Vol. II, p. 709; and see J.S. Hill (ed.), *Imagination in Coleridge*, London, 1978, Macmillan, p. 10.

9 Herder, quoted by Roy Pascal, p. 183.

10 Charles Taylor, *Hegel*, Cambridge, 1975, CUP, p. 15.

11 See Chapter 7.

12 Cited by Charles Taylor, *The Sources of the Self*, Cambridge, 1989, CUP, p. 371.

13 See, for example, Alasdair MacIntyre, *A Short History of Ethics*, London, 1966, Routledge & Kegan Paul, p. 39.

14 See Charles Taylor, *Sources of the Self*, Chapter 23.

15 Rudolf Otto, *The Idea of the Holy*, trans. by J.W. Harvey, 1917, Pelican edn, Harmondsworth, 1959, Penguin Books.

16 From the Brihadaranayaka Upanishad, quoted by W.T. Stace in *Mysticism and Philosophy*, London, 1961, Macmillan, p. 118.

17 Meister Eckhart, quoted by Stace, ibid., p. 63.

4 The exploitation of nature and women

1 Val Plumwood, *Feminism and the Mastery of Nature*, London, 1993, Routledge.

2 See Stephen Clark, 'Platonism and the gods of place', in *The Philosophy of the Environment*, T.D.J. Chappell (ed.), Edinburgh, 1997, Edinburgh University Press.

3 Plumwood, *Feminism and the Mastery of Nature*, above, p. 3.

4 Plumwood, above, p. 104.

5 Plumwood, above, p. 4.

6 Where the gender dimension is not obvious, there is always a cultural assumption that makes the link. Plumwood, above, p. 45.

7 Quoted by Carolyn Merchant, *Radical Ecology*, London, 1992, Routledge, p. 276.

8 Merchant, above, p. 191.

9 Merchant, above, p. 194.

10 See Murray Bookchin, *The Philosophy of Social Ecology*, Montreal, 1990, Black Rose.

5 Phenomenology and the environment

1   H.I. Dreyfus and S.E. Dreyfus, 'What is morality? A phenomenological account of the development of ethical expertise', in D. Rasmussen (ed.), *Universalism vs. Communitarianism*, Cambridge, MA, 1990, MIT Press. This is on morality rather than environment; but it is a very interesting example of phenomenology at work.

6 Coping with individualism

1   Karl Marx, *Theses on Feuerbach*, VI, available for example in Marx and Engels, *Basic Writings on Politics and Philosophy*, ed. Lewis S. Feuer, London, 1959, Fontana.
2   Alasdair MacIntyre, *A Short History of Ethics*, London, 1966, Routledge & Kegan Paul, p. 124.
3   David Hume, *A Treatise of Human Nature*, London, 1739, Book III, Part III, §1.
4   See Herbert Marcuse, *One Dimensional Man*, 1964; second edn, London, 1991, Routledge.
5   Hobbes, *Leviathan*, 1651, Chapter XIV.
6   Adam Smith, *An Inquiry into the Nature and Causes of the Wealth of Nations*, London, 1776.
7   Garret Hardin, 'Tragedy of the commons', 1968, reprinted in *Environmental Ethics*, eds R.G. Botzler and S.J. Armstrong, Boston, MA, 2nd edn 1998, McGraw-Hill, pp. 520–3.
8   See John O'Neill, *Markets*, London, 1998, Routledge.
9   John Locke, *Second Treatise on Government*, London, 1690, Section 6, Chapter 2.
10 Tom Regan, *The Case for Animal Rights*, Berkeley, 1983, University of California Press, p. 243.

7 Lines into the future

1   'Wherever there is a conscious mind, there is a point of view. This is one of the most fundamental ideas we have about minds – or about consciousness.' Daniel Dennett, *Consciousness Explained*, 1991, Harmondsworth, Penguin edn, 1993, p. 101.
2   Gilbert White, *The Antiquities of Selborne*, Bowdler Sharpe edn, *The Natural History and Antiquities of Selbourne and a Garden Calendar*, 1789, ed. R. Bowdler Sharpe in 2 vols, London, 1900, Freemantle, p. 315.
3   See R.B. Braithwaite, *Scientific Explanation*, Cambridge, 1953, CUP.
4   An excellent exploration is Andrew Woodfield, *Teleology*, Cambridge, 1976, CUP.
5   Paul Taylor, 'The ethics of respect for nature', *Environmental Ethics* 3: 210.
6   Taylor, above, p. 211.
7   Taylor, above, p. 207, my emphasis.
8   There is an extended discussion of the rules of conduct in Chapter 4 of Taylor's book, *Respect for Nature* (Princeton, NJ, 1986, Princeton

University Press). A useful brief account of them, with some discussion, is given in Joseph R. DesJardins, *Environmental Ethics*, Belmont, CA, 1993, Wadsworth, pp. 157–62.

9 Taylor, above, p. 215.

8 Ecology and communities

1 I draw heavily on Donald Worster, *Nature's Economy – a history of ecological ideas*, Cambridge, 1977, CUP.
2 What determined the pattern characteristic of a particular place was climate. Deciduous forests were another 'formation', prairies another.
3 This was the contribution of the Dane Eugenius Warming, whose seminal statement, *Plantesamfund*, was published in 1895. The quoted phrase is Warming's, cited by Worster, above, p. 199.
4 By Anton de Bary in 1879.
5 Worster, above, p. 210. I say 'apparently' because of the tension between stressing the reality of the climax formation characteristic of a particular region, and insisting on *change* as the keynote.
6 Worster, above, p. 204.
7 Vernon Pratt, *Thinking Machines*, Oxford, 1987, Blackwell, Chapter 13.
8 Others took this forward by focusing on energy flows between the parts of a mechanical system such as this. Still others made much of the industrial model and spoke of products, productivity, yield, efficiency and investment.
9 For an accessible outline of ecological ideas and their significance to ethics see Holmes Rolston III, 'In defence of ecosystems', *Garden* 12, 1988.
10 Aldo Leopold, *The Sand County Almanac*, 1949, Oxford, 1977 edn, OUP.
11 See Chapter 6. I myself am so enmired in individualism that I confess this line of thought makes me not a little queasy…
12 See the discussion of Hume's theory of ethical behaviour in Chapter 6, above.

9 The importance of being an individual

1 See Chapter 1.
2 This was Leibniz's thesis.
3 See the attack on 'reductionism' by Romantic thought, indicated in Chapter 3.
4 Roger Woolhouse, *Descartes, Spinoza, Leibniz*, London, 1993, Routledge, p. 10.

10 The aesthetics of the natural environment

1 In this chapter I use the term 'environmental aesthetics' as a description of aesthetics of the natural environment. In the wider context of aesthetic theory it refers to aesthetic experience and value of both the

built and natural environments. For example, it might also refer to our aesthetic encounters with the urban environment.

2 Kant, *Critique of Judgment*, trans. W. Pluhar, Indianapolis, 1987, Hackett.

3 For Kant, the judgement of taste is an expression of finding an object to be beautiful. He also claims that when we feel displeasure in relation to the perceptual qualities of an object, we will find the object ugly. I do not consider this point in any depth here because Kant himself had very little to say about it.

4 Kant, *Critique of Judgment*, above, §1.

5 Kant, *Critique of Judgment*, above, §9.

6 Kant, *Critique of Judgment*, above, §8.

7 Kant, *Critique of Judgment*, above, §6, §7.

8 Kant, *Critique of Judgment*, above, §2, §5.

9 This point rests not only on Kant's arguments in the *Critique of Judgment* (cf. §§ 20–2), but also many of the arguments of his first critique, the *Critique of Pure Reason* (trans. Norman Kemp Smith, London, 1929, Macmillan), where he sets out the conditions for the possibility of objective experience and knowledge. Although he distinguishes aesthetic judgement from cognitive judgement, aesthetic judgement is dependent upon the very same perceptual and mental apparatus that makes objective experience possible. From this he concludes that aesthetic judgements retain the shared foundation of cognitive judgement, despite the fact that they are grounded in feeling rather than concepts.

10 Kant, *Critique of Judgment*, above, §45.

11 The differences I set out here have been recognised by various philosophers, but most notably by Hepburn, 'Nature in the light of art' and 'Contemporary aesthetics and the neglect of natural beauty'; Y. Sepänmaa, *The Beauty of Environment: A General Model for Environmental Aesthetics,* Denton, 1992, Environmental Ethics Books; and Carlson, 'Environmental aesthetics', in D. Cooper (ed.), *A Companion to Aesthetics*, Oxford, 1992, Blackwell, and 'Appreciating art and appreciating nature', in S. Kemal and I. Gaskell (eds), *Landscape, Natural Beauty and the Arts*, Cambridge, 1993, Cambridge University Press. For the role of design, see especially Carlson, 1993, above.

12 I should also note that there is a sense in which even artefacts have a natural origin, in so far as the humans who produce artefacts are themselves part of nature.

13 This point will of course not hold for individuals who regard nature as the product of divine creation. But the point is still relevant for distinguishing between art as the product of human creativity and imagination, and nature as something that is not created as a work of art.

14 Many artistic conventions have been challenged in this century, so that it is becoming more common to experience art not as an object set apart from us, as is so typically the case with paintings or theatre. Contemporary art has challenged these conventions through art that encourages participation, such as certain art installations, as well as theatre productions that require audience participation. In these cases it is more possible to experience art as an *environment*.

15 Hepburn, 'Contemporary aesthetics and the neglect of natural beauty', in *Wonder and other Essays*, Edinburgh, 1984, Edinburgh University Press, p. 13.
16 Hepburn, 'Nature in the light of art', in *Wonder and other Essays*, Edinburgh, 1984, Edinburgh University Press, p. 51.
17 Hepburn, 'Contemporary aesthetics and the neglect of natural beauty', Note 15 above, p. 19.
18 Walton 1970. The example is my own.
19 This model emerges from the fashion for the picturesque, an eighteenth-century aesthetic category that encouraged viewing nature as a painter might in the process of painting a picture of it.
20 For a discussion of these criticisms of the landscape model, see Hepburn, 'Aesthetic appreciation of nature', in H. Osborne (ed.), *Aesthetics in the Modern World*, London, 1968, Thames & Hudson; and 'Contemporary aesthetics and the neglect of natural beauty', in *Wonder and other Essays*, Edinburgh, 1984, Edinburgh University Press.
21 It might be argued that design is never an appropriate category of aesthetic appreciation of nature, but I think there will be cases in which it is. One might believe that the complex design of a snowflake is the design of evolutionary causes, or is caused by a divine designer. Kant recognised that when we find beauty in a natural object, part of this judgement stems from feeling that it is as if the object has been designed for our (disinterested) pleasure. So in these ways design may be a relevant aesthetic feature in appreciation. However, it will not be relevant if design is taken to mean designed by an artist, and it is counter-intuitive to think that this could ever be appropriate to nature. A different sort of case occurs when we say, for instance, that some aspect of the countryside 'looks just like a Constable picture'. Here it is merely our experience of artistic culture that is brought to bear on what we see, such that we see the countryside as a Constable picture. This case is different from framing appreciation of the countryside with a picture frame, as in the landscape model. It is not intentional, nor is it restrictive on appreciation, but rather it opens a connection between culture and nature.
22 Examples of the science-based model include Carlson, 'Appreciating art and appreciating nature', in S. Kemal and I. Gaskell (eds), *Landscape, Natural Beauty and the Arts*, Cambridge, 1993, Cambridge University Press; M.M. Eaton, *Aesthetics and the Good Life*, Rutherford, NJ, 1989, Farleigh Dickinson University Press; Rolston, 'Does aesthetic appreciation of landscapes need to be science based?', *British Journal of Aesthetics* 35, 1995, pp. 374–86.
23 The two central figures in the non-science-based approach are Berleant and Hepburn. Other notable work includes: von Bonsdorff, 'Paces of change', in P. von Bonsdorff (ed.), *Ymparistoestiikan Polkuja (Paths of Environmental Aesthetics)*, International Institute of Applied Aesthetics Series, Vol. 2, 1966, Jyväskylä, Gummerus Kirjapaino Oy; Carroll, 'On being moved by nature: Between religion and natural history', in S. Kemal and I. Gaskell (eds), *Landscape, Natural Beauty and the Arts*, Cambridge, 1993, Cambridge University Press; C. Foster, 'Aesthetics of the natural environment', unpublished Ph.D. dissertation, University of Edinburgh, 1992; S. Godlovitch, 'Icebreakers, environmentalism and natural aesthetics', *Journal of Applied Philosophy*, 1994, pp. 15–30. See also Sepänmaa, *The Beauty of Environment: A General Model for Environmental Aesthetics*, Denton, 1992, Environmental Ethics Books,

for an overview of the problems in constructing an aesthetics of the natural environment.

24 Carlson, 'Nature and positive aesthetics', *Environmental Ethics* 6, 1984, p. 26.

25 Carlson, 'Nature and positive aesthetics', above, p. 25.

26 Carlson, 'Appreciating art and appreciating nature', in S. Kemal and I. Gaskell (eds.), *Landscape, Natural Beauty and the Arts*, Cambridge, 1993, Cambridge University Press, p. 218.

27 Carlson, 'Nature and positive aesthetics', *Environmental Ethics* 6, 1984, p. 26.

28 See Berleant, *Aesthetics of the Environment*, Philadelphia, 1992, Temple University Press.

29 Rolston, 'Does aesthetic appreciation of nature need to be science based?', *British Journal of Aesthetics* 35, 1995, pp. 374–86, and M.M. Eaton, 'The role of aesthetics in designing sustainable landscapes', in Y. Sepänmaa (ed.), *Real World Design: The Foundation and Practice of Environmental Aesthetics*, Proceedings of the XIIIth International Congress of Aesthetics, Vol. II, 1995.

30 Hepburn, 'Contemporary aesthetics and the neglect of natural beauty', in *Wonder and other Essays*, Edinburgh, Edinburgh University Press, 1984.

31 Hepburn, 'Contemporary aesthetics and the neglect of natural beauty', above, p. 27.

32 Ibid.

33 Hepburn, 1993, p. 69.

34 Berleant's position is the most deeply phenomenological of these views (Berleant, *Aesthetics of the Environment*, above), but the influence is also significant in von Bonsdorff, 'Paces of change', above; and Hepburn, 'Contemporary aesthetics and the neglect of natural beauty', above.

35 Berleant, *Aesthetics of the Environment*, above, pp. 236–7.

36 Risto Lotvonen, resident of Hyvinkää, quoted in von Bonsdorff, 'Paces of change', above, p. 130.

37 Carroll, 'On being moved by nature: Between religion and natural history', pp. 257–8.

38 Carlson, 'Appreciating art and appreciating nature', above, p. 226.

39 Rolston, 'Does aesthetic appreciation of nature need to be science based?', p. 376.

Index